Raising Goats

2nd Edition

by Cheryl K. Smith

Raising Goats For Dummies®, 2nd Edition

Published by: **John Wiley & Sons, Inc.,** 111 River Street, Hoboken, NJ 07030-5774, www.wiley.com

For general information on our other products and services, please contact our Customer Care Department within the U.S. at 877-762-2974, outside the U.S. at 317-572-3993, or fax 317-572-4002. For technical support, please visit https://hub.wiley.com/community/support/dummies.

Wiley publishes in a variety of print and electronic formats and by print-on-demand. Some material included with standard print versions of this book may not be included in e-books or in print-on-demand. If this book refers to media such as a CD or DVD that is not included in the version you purchased, you may download this material at http://booksupport.wiley.com. For more information about Wiley products, visit www.wiley.com.

Library of Congress Control Number: 2021930797

ISBN 978-1-119-77258-3 (pbk); ISBN 978-1-119-77259-0 (ebk); ISBN 978-1-119-77260-6 (ebk)

Manufactured in the United States of America

SKY10024694_020421

Contents at a Glance

Table of Contents

PART 2: BRINGING YOUR GOATS HOME 73

Introduction

Welcome to *Raising Goats For Dummies.* Raising goats is a rewarding and challenging adventure. Whether you are new to it or have been raising goats for a few years, you never stop finding out about or being delighted by these intelligent creatures.

I wrote this book to help you get started successfully raising goats and to answer any questions that you have. Whether you're raising goats for fun, for companionship, or to add value to your farm and your lifestyle, this book gets you on the right track.

About This Book

I took ten years between thinking about getting some goats and actually getting them. I had to get a place where I could keep them and then try to figure out what I wanted and what I was getting into. After I actually became a goat owner, I had to figure out a lot through trial and error.

I needed a book that laid out the basics, gave me tips on what to look for, what to expect, and how to avoid some of the most common pitfalls. This book does that. It gives you the basic information you need to make the intelligent decisions required to successfully raise goats and helps you avoid making mistakes that have a negative effect on their lives and yours.

Even if you've been raising goats for a few years, this book can help you. You find new ideas on how to keep your goats healthy and happy. And you find out how you can use the goats you have to become more self-sufficient. This book is for anyone who wants to raise goats.

Within this book, you may note that some web addresses break across two lines of text. If you're reading this book in print and want to visit one of these web pages, simply key in the web address exactly as it's noted in the text, pretending as though the line break doesn't exist. If you're reading this as an e-book, you've got it easy — just click the web address to be taken directly to the web page.

Foolish Assumptions

When writing this book, I made some assumptions about who you, the reader, might be. I assume that you

- Are already planning to get some goats or have a few and want to find out more about caring for them
- Are interested in the basics of raising goats for milk, meat, fiber, backpacking, or pets
- Want to produce some of your own food so you can control the quality and what goes into it
- Don't know much about goat health care and plan to work with a veterinarian when your goats get sick
- Want to save money by building some of the structures and supplies your goats need

Beyond the Book

In addition to the material in the print or e-book you're reading right now, this product also comes with some access-anywhere goodies on the web. Check out the free Cheat Sheet for tips on preparing your property for goats, the signs of a sick goat, and the questions to ask before buying a goat. To access the Cheat Sheet, go to `www.dummies.com` and type **Raising Goats For Dummies Cheat Sheet** in the Search box.

Icons Used in the Book

Throughout this book, little pictures in the margins draw your attention to special types of information that make your reading experience more helpful. Here's what you find:

REMEMBER

Some information bears repeating, and I highlight it with this icon. Important points that you may turn to again and again appear next to it.

TECHNICAL STUFF

The information you find next to this icon is more detailed than usual. When you see it, feel free to move on. You won't lose out on main points.

TIP

Beside this icon are nuggets of important information that help you to be a better goatkeeper. I use this icon to show you ways to save time or money — or both.

WARNING

This dangerous-looking icon draws your attention to potential bad outcomes or mistakes that you want to avoid. Pay close attention to them.

Where to Go from Here

Pick a chapter, any chapter. Each one is its own little book. You won't need to go back to fill in missing pieces from earlier chapters. Looking for information about what to expect from kidding? Turn to Chapter 13. Want to glimpse the details of caring for goats? Chapter 9 has what you need. And if you're an overachiever or just insatiably curious, by all means turn the page and keep going until you get to the back cover.

I imagine that the more you find out about goats, the more likely you are to fall in love with them. They're smart, curious, and calming, and they can even help you make some money. So welcome to the world of goats. I'm grateful for the opportunity to help you on your quest.

1

Getting Started with Goats

IN THIS PART . . .

Discover the benefits of raising goats and decide whether goat keeping is for you.

See what makes a goat a goat.

Discover the pros and cons of different breeds and figure out which breed is right for you.

Ensure that your goats have a happy, healthy life by getting your property goat-ready.

IN THIS CHAPTER

» Finding out about the many benefits of raising goats

» Getting to know these lovable animals

» Deciding whether goat keeping is for you

Chapter 1

Discovering the Joys of Raising Goats

I've been raising goats since 1998, and I'm still not tired of them. In fact, I can't imagine my life without them. Talk to any other serious goat keepers and you hear that after they get "in your blood" you have to have at least a couple.

When you get a sense of goats' many uses and get to know their unique qualities, goats will get in your blood, too. Goats are more than livestock, they're friends and helpmates, and they're entertaining as heck.

In fact, the popularity of goats, especially dairy goats, has grown in the United States in the past decade. That doesn't mean that the same people now have bigger herds. In fact, the average size of a goat herd went from a little over 29 heads in 2009 to almost 20 heads in 2019. According to the U.S. Department of Agriculture (USDA) National Animal Health Monitoring System (NAHMS) Goat Study, small herds of 5 to 19 goats are not just pets; more than half of all dairy and meat goat farms and ranches have small herds. Dairy goats are becoming even more popular, increasing 57 percent in just ten years, but almost one-quarter of goat owners surveyed said they own their goats for more than one reason, such as milk, brush clearing, packing, or as pets.

Goats have been called the poor man's cow but a better name for them would be the "green" cow. They take up less space, cost less to feed, and even contribute less to global warming because they don't emit as much methane gas. On top of all that, they're smart, friendly, and easy to manage with the proper setup. This chapter introduces you to the benefits of raising goats and lets you know what you can expect from the experience.

Finding Goat Basics

Goats are intelligent and friendly animals. They come in all sizes (but only one shape) and can help you in numerous ways. After you decide what you want to accomplish with them, you have a lot of breeds to choose from. (Most goat owners have their favorite, which they can talk about endlessly.) Check out Chapter 3 for a rundown of goat breeds.

Goats are unique animals within the ruminant classification. *Ruminants* are also known as "cud-chewing" animals, or as having four stomachs. You can find out more about their digestive systems in Chapter 2, as well as how to tell them from sheep, what makes a healthy goat, and how goats communicate with you and each other.

Being responsible for goats is a serious undertaking. I say "goats" because goats need a herd to be happy, so you shouldn't get just one goat. You can read more about why you need at least two goats and other ways to keep them happy in Part 2.

THANK GOATS FOR COFFEE

According to legend, goats in the charge of an Ethiopian goatherd named Kaldi failed to return one evening. When he found them the next morning, they were excited and dancing next to a shrub covered in red berries. He investigated, trying the berries, and discovered that they were responsible for the goats' excitement. The goats had discovered coffee.

When the goatherd took the berries to the town monastery, the abbot disapproved and threw them into the fire. When he smelled the pleasant odor they gave out as they roasted, he raked them out of the fire, ground them up, and mixed them with water. Coffee's use as a stimulant drink gradually spread around the world.

Goats still enjoy coffee, and some goat owners use it to stimulate labor and give energy to a *doe* (female) that is kidding.

These critters can live for ten years or more, so getting goats is like getting a dog — you may be taking care of them for quite a while. Chapter 2 tells you about their life expectancy.

Identifying the Benefits of Owning Goats

Goats are fantastic animals that have been domesticated for more than 10,000 years. You get a lot from a relatively small animal — you can milk them or eat their meat, use their fiber and their skin for making clothing, and even use their dung for fuel (if you are so inclined). In the past, goat hide was made into bags for carrying water and wine and parchment for writing on; it is still used to make drums in some countries.

You may want to raise goats for a variety of reasons. Whatever brings you to goats, you're guaranteed to find additional benefits to owning these critters after you start working with them.

Becoming more self-sufficient

Goats are not only a great way to become more self-sufficient — they can give you milk to drink and food to eat, and even help you carry your belongings when backpacking — but they teach you in a very direct way where your food comes from and give you an opportunity to affect its quality. If prices go up (and don't they always?) you are less affected if you're supplying some of your own food.

And imagine not ever having to cut down blackberries or kudzu again. You can get your friendly goat to do it for you, while growing fiber for hats and sweaters and providing you with milk, meat, and even more goats.

In this section I talk about the many ways goats can contribute and move you toward self-sufficiency.

Cutting your dairy bill

Imagine never having to buy milk or cheese again. If you raise dairy goats you can achieve that goal. Your goats need to have kids to give you milk, and then you can milk them throughout the year for up to five years without rebreeding, if you want. Or you can stagger the kidding each year so that you have a milk supply year-round. (They need a break from milking during the last two months of their pregnancy to put their energy into growing kids.)

You need only a few goats to keep a small family in milk and other dairy products. Just one standard-size dairy goat can give you an average of 6 to 8 pounds (3 to 4 quarts) of milk each day. And, depending on the butterfat content of the milk, you can get up to a pound of cheese for every gallon of milk.

With your own milk supply, you won't need to worry about additives to the milk you drink or the cheese or yogurt you eat. Goat milk is easier to digest than cow milk, and so it is prized by people who can't drink cow milk. Depending on the state you live in, you can sell milk to supplement your income or offset the goats' feed costs.

You find out about raising goats for milk in Chapter 15.

Raising your own meat

Goat meat has always been popular in the developing world, because goats are much more affordable and use fewer resources than animals such as cows. According to the U.S. Department of Agriculture, the demand for goat meat is expected to continue growing.

People who moved to the United States from Latin America, the Caribbean, the Middle East, Asia, and Africa brought their custom of eating goat meat as a regular part of their diet and still want it. And more people who hadn't eaten goat meat before are willing to try a new, lean source of protein that doesn't have the taint of confined animal feeding operations (CAFOs). Animals raised in CAFOs often never see the light of day and are unable to exercise or eat grass because they are packed into small areas.

DRINKING MILK ON THE FARM MAY HELP YOU BREATHE EASIER

A number of studies have shown a positive correlation between living on a farm and *not* having allergies or asthma. One study goes even further, showing a likely benefit of raising goats for their milk.

A 2017 study of farmers and their spouses in the United States looked at correlation between a tendency to develop allergies and exposures to a farming environment (including early childhood farm animal contact and raw milk consumption). The results backed up earlier studies from Europe showing a decreased tendency to develop allergies when in a farming environment. This was especially true in cases where their mothers had performed farm activities while pregnant, as well as when they drank raw milk and were exposed to farm animals before the age of 6.

Goat meat is easily digestible, tasty, and low in fat. If you're in charge of your own source of meat, you know how the animal was raised and what feed or medications went into the animal.

Meat goat farming provides a great opportunity to start a business or supplement your income. You can raise goats that are bred for meat, or you can use your excess dairy *bucks* (males) or *wethers* (bucks that have been castrated) as an alternative to buying meat or to provide income to support your herd. Before you jump into a meat-goat enterprise thinking that you're going to get rich, you need to investigate a few things:

- **Check out the market for goat meat in the area of the country where you live.** Contact your extension office for assistance, go to the local livestock auction to see how well and for what price meat goats are selling, and read the local agriculture newspaper or other publications to see what they have to say about raising and marketing meat goats.
- **Learn about *stocking* (the number of animals you can support without overcrowding or overgrazing), how many goats you can stock on your property, and how that might break down in terms of income.**
- **Determine what kind of meat goats are available and for what price.**
- **Consider slaughter options.** Would you slaughter goats yourself, sell them at auction, have a mobile slaughter provider come out, or transport your goats to a slaughterhouse? Do you have a vehicle to transport goats? Is there a USDA-certified slaughter facility nearby that handles goats? Factors affecting this decision include laws governing slaughter as well as the local market, your capabilities, and financial considerations.

I talk in much more detail about raising goats for meat in Chapter 16.

Growing your own fiber

Some of the finest fiber comes from goats: Angora and Pygora goats produce mohair, cashmere goats produce cashmere, and crosses between the two breeds produce a fiber called *cashgora.* An adult angora goat can produce an average of 8 to 16 pounds of mohair each year, and a kid can produce 3 to 5 pounds. Cashmere and cashgora-producing goats produce less fiber, but it is also more highly valued.

If you raise fiber goats, you can spin your own yarn and make hats, blankets, sweaters, or other products. You can also sell the fiber to spinners or to companies that make these products, while having the benefit of these friendly creatures.

Check out Chapter 18 to find out about harvesting and using goat fiber.

Harnessing goats' power as living weed whackers

Goats are well-known for their ability to wipe out weeds. In fact, some people have made businesses out of renting out their goat herds to cities and other municipalities to clean up areas that are overgrown with weeds or blackberry bushes. These leased goats decrease the need to use herbicides, improve the soil's fertility, decrease the risk of fire, increase the diversity of plants in the area, and control weeds in hard-to-reach areas, such as steep hills.

TIP

Because goats are browsers, they can share or alternate a pasture with sheep or cattle, which prefer different plants. Goats eat brush, leaves, and rough plants. They can improve pasture by removing noxious weeds, clear areas to be replanted with trees, and control leafy spurge, knapweed, Himalayan blackberry, giant ragweed, sunflowers, kudzu, and other weeds.

WARNING

Not every plant is a great snack for a goat. I tell you about plants you need to keep away from your goats in Chapter 4.

Whether your goats are pets, milk producers, meat animals, or serve another purpose, they provide the side benefit of acting as living weed whackers. With some portable fencing or a guardian animal for protection, they range far and wide each day to keep your property free of noxious weeds.

Don't expect to put them on a lawn and have them mow down the grass, though. "Lawnmower" is the job of sheep, not goats. Goats prefer to eat your rosebushes or lilacs.

CASHMERE GOATS

No specific breed of goat is named *cashmere*. However, feral goats from Australia and New Zealand and Spanish meat goats from the southwestern United States can be registered with the Cashmere Goat Association, if they meet the standard. Breeders have produced more productive cashmere goats by selectively breeding good producers from these populations. Cashmeres have the added benefit of being good meat goats.

The term *cashmere* refers to the undercoat or down that is harvested from a variety of goats. Cashmere is harvested and processed mainly in central Asia, especially China. The fiber produced by up to four of these Asian feral goats in a whole year is required to make just one cashmere sweater — which explains why they are so expensive.

Breeding and selling

Unless your goats are just pets or brush eaters, you probably want to breed them. If you have dairy goats, you need to breed them to keep a good supply of milk flowing. And you will need to replace any goats you sell or slaughter.

As a dairy goat owner, I supplement my income from selling milk by selling kids and providing buck service. *Buck service* means leasing a buck for breeding purposes to another goat owner. Buck service is valuable to goat owners who don't have the space or don't want the hassle of keeping a buck or who want to get certain genetics into their herd. I tell you more about buck service, and about breeding in general, in Chapter 12.

Using goats for companions or helpers

Goats make great companions, something that more people discover every day. Miniature goats such as the Nigerian Dwarf, Pygmy, and mini dairy breeds are growing in popularity as pets, in both the city and the country. (Check out Chapter 3 to find out more about these and other breeds.)

Goats are intelligent and funny, and they're also a great way to meet people. I had a little goat named Malakai, who was a *dwarf* Nigerian Dwarf, because of health issues. His petite size made him all the more adorable. He was house- and car-broken, and so I took him with me wherever I went. He was the little Mystic Acres farm emissary and never failed to attract people. Besides helping me meet people, Malakai gave me the opportunity to educate people about goats and clarify their inevitable misunderstandings. (Get into goats, and you'll find that misconceptions abound. I brace you to take on the most common of them in Chapter 20.)

Keeping goats as pets

REMEMBER

Normally considered livestock, goats can make good pets, but you will be much more satisfied using them for this purpose if you remember that they are outdoor animals and that's where they do best. Goats are herd animals and need another goat for a friend, so get at least two of them. I tell you more about choosing your goats in Chapter 5.

You can leash train goats and take them on walks throughout the neighborhood or around your property, which provides exercise for all of you. (Chapter 8 shows you how to get started leash-training.) I can tell you from experience that just sitting and watching goats has a calming effect. Studies of other pets have shown that they can lower your heart rate and improve your health — and I'm sure that the same is true for goats as pets.

Finding a helping hoof: Using your goat for packing

Goats are social animals and, after you establish a relationship with them, they love to spend time with you. They enjoy going for hikes and can go almost anywhere you can. Not only that, but they can carry your belongings, they find plenty to eat right there in the wilderness, and they make great companions.

So why not take your goat packing? Ideally, you select a large wether for packing and then take the time to train him to obey commands and to carry your gear. Goats are surefooted, excellent pack animals and can help you work, whether it be gathering wood in the forest or just carrying your belongings on a holiday hike. Chapter 8 tells you about training your goat as a pack animal.

Sharing your goat with others

Another way you can use a companion or pet goat is as a therapy animal or a visitor to children, seniors, or other groups who don't usually see goats. After you train your goats, you can approach teachers, administrators, or activity directors to plan a goat day (or hour). You get to spend time with your goats, and other people get to learn about their unique personalities.

Some people use goats to help children with autism improve their sensory abilities and social skills — there's nothing like a cute pet to get people talking to you — and to improve morale and entertain residents of nursing homes. (I talk more about these benefits in Chapter 18.)

Raising goats as a 4-H project

Getting children involved in raising goats is a good way to teach responsibility. Keeping goats requires twice-a-day chores. Children quickly learn that the goats depend on them. They also find out about the cycle of birth and death and get outdoors to get regular exercise.

TIP

Learning about and caring for goats as a 4-H project provides a structure that makes caring for goats fun and easy by giving the project a bigger purpose. Contact your county extension office for help on finding a 4-H group. If the 4-H group in your area is not set up so your kids can raise goats for a project, consider getting trained and starting your own goat 4-H group. Doing so not only gives your kids the opportunity for such a project, but it teaches responsibility, helps promote goats, and educates other children.

Some of the things that children can do in a goat 4-H project include the following:

- Developing a budget for goat care
- Writing a report on and giving a speech about goats
- Demonstrating hoof-trimming or other routine care (Chapter 9)
- Watching a goat show
- Exhibiting the goat at the county or state fair (Chapter 17)
- Milking the goat and making cheese or yogurt (Chapter 15)
- Training the goat to walk on a lead (Chapter 8)
- Writing or drawing for a goat newsletter or magazine

Determining Whether Goats Are for You

You may love goats and the idea of raising them, but how do you know whether raising goats is right for you?

The first order of business is finding out everything you can about goats. You get a good start by reading this book. I also recommend that you spend time around goats. Ask goat people you've met whether you can go to their farms and observe or even help with their goats. Firsthand experience gives you an idea of what goats need and how you like working with them.

REMEMBER

Goats can live 15 years, or even longer. Unless you buy goats to be eaten, and especially if you plan to keep them as pets, remember that you're taking on a long-term commitment just like you do when you get a dog or a cat.

The upcoming sections tell you more about considerations you need to mull over before you become a goat owner.

Devoting time and effort

Expect to spend at least a half hour each morning and a half hour each evening on routine goat care. If you get a lot of goats or use them for a specialty such as milk, meat, or fiber, you need to budget more time. (Part 4 tells you about each of these situations.)

Goats need a supply of hay and/or *browse* (grasses and other plants they can find in a pasture) and clean water at all times. (Chapter 6 gives you the details on what goats eat.) The routine twice-daily care you need to plan for includes feeding, changing water, cleaning buckets, observing your goats to make sure they are healthy and acting normal, making sure they're safe and secure each night, and letting them out in the morning. Of course, you probably want to spend more time just being with them after you discover how fascinating they are.

Plan also to spend an hour or more each month on regular grooming and goat care such as hoof trimming, injections or other treatment, and cleaning their living area. (Chapter 9 runs through the routine care that goats need.) If you breed, show, shear, or slaughter your goats, you spend many more hours with your goats intermittently. And if one of your goats gets sick, you need time to provide care or coordinate with a vet. (Chapter 11 addresses common illnesses.)

REMEMBER

If you work and don't have a reliable helper, you need to have some flexibility to deal with problems. If you work outside your home and have long or erratic hours, a helper is essential.

Deciding which goats are right for your situation

You want to get the type of goat that's right for your goals. If you want goats for milk, get dairy goats; for meat, get meat goats; for fiber, get fiber goats. If you have mixed goals, find out which animal will best suit all of them. For example, a Spanish goat can produce cashmere, kids for meat, and milk. The milk won't be of the volume or quality you get with a dairy goat, but nevertheless, it's milk. If you just want pets, miniature goats may be your best bet. And for brush control, bigger may be better. You can read more about the different breeds in Chapter 3.

REMEMBER

If you want pet goats, choose goats that are

- **Tame and friendly:** These goats are much easier to work with, and you won't have to spend time trying to get them to trust you.
- **Horn-free:** Goats without horns are the safest, especially around children. (I tell you more about the horns, including how to remove them before they grow, in Chapter 9.)
- **Wethers:** Don't ever accept or buy a buck goat for a pet. You will regret it as soon as he matures and starts to stink during breeding season. (Chapter 12 tells you about the weird and exciting world of breeding.) Does can make okay pets, but wethers are the best. They don't go into heat, and they cost less to feed because they don't need anything but minerals and good grass hay. They also are the sweetest.

If you live in a city, get miniature goats, not full-sized ones. They won't take up as much room and then you can have more, if you want. By the same token, if you have a physical disability, miniature goats are better because they are easier to handle. You can read more about miniature goat breeds in Chapter 3.

If you'd like to raise and breed show goats, you probably want to get registered goats. There are many registries for goats, usually related to their use — for example, dairy goats, meat goats, or cashmere goats. I tell you more about showing goats in Chapter 17.

Finding out about local ordinances

Check out ordinances in your area regarding keeping livestock. These govern whether you can keep livestock and restrict the types you can keep. You need to be aware of the ordinances governing your area, or you risk losing the goats that you dreamed of getting just because of a technicality.

TIP

A good place to start is with a city planning or zoning department. Make sure to get a copy of the rules or regulations that govern this. Jenny Grant, of the Seattle Justice League, who spearheaded the Seattle effort to allow miniature goats in that city, also recommends that you check with local animal control to determine whether they have the ability to deal with goats.

In some cities, you may need to buy a license for a goat, just as you do for a dog. Another kind of ordinance that you need to be aware of governs noise. In rural areas, animals are expected (as are guns), so neighbors can't do much legally about a crying goat (or target shooting). But in an urban area, even if goats are allowed, your neighbors may complain, much as they do with a barking dog. Be aware of what your local noise ordinance covers.

In some cases where goats are prohibited by a city or other municipality, people with special needs or medical issues have been able to obtain variances to the zoning so that a goat could serve as a service or therapy animal.

TIP

If your city does not allow backyard goats, you'll need to work with the powers that be to get the law revised, like other urban farmers are doing in cities around the United States. Grant suggests putting together a petition and obtaining signatures and then taking it to the city council and requesting a hearing. If you also obtain email addresses on the petition, you'll have a ready-made list to notify interested parties when the hearing is scheduled.

Also helpful are recommendations on what should be covered by such an ordinance, such as miniature goats only, a limitation on the number, only does or wethers, and whether they need to be licensed. You can use another city ordinance as a pattern for the one you're proposing.

Knowing your neighbors

Even more important is to know your neighbors. Do they have any vicious or out-of-control dogs that may be a threat to your goats? Can you keep these dogs away from your goats?

Do you already have bad blood with neighbors? Will they bitterly complain, undermine you, or otherwise make your life miserable if they hear goat cries? If you live in an area where you'll need a livestock guardian dog, will neighbors have a problem with nighttime barking? Try to work things out first, but if you can't, think about whether you want to invest a lot of time, money, and heart in a project where a nearby neighbor will make your life difficult.

TIP

After you get goats, share some of your goat products with your neighbors, consider involving them or their children in your project, and work with them to make sure your goats aren't causing a problem.

Better yet, if they're interested, teach your neighbors about goats. You could even offer to work with a neighbor child who is interested in raising goats for 4-H. If you have neighbors who have enough interest, they may be willing to learn enough to be able to care for them so you can take a vacation. It's a win-win!

» Focusing on features that make a goat a goat

» Identifying a healthy goat

Chapter 2
Glimpsing Vital Goat Statistics

Goats are interesting and amazing animals. They're related to sheep, deer, and cattle but have some basic differences. Talking about goats and their various parts requires you to pick up some new vocabulary.

In this chapter, I give you an introduction to goat terminology, take you through body parts that are integral to "goatness," and give you some tips on what to expect from a healthy goat and how to tell whether things are starting to go wrong.

Doe, a Goat, a Female Goat

If you want to get goats, you need to talk goats with other goat lovers. To help you avoid missteps, here is the basic terminology that you need to ask questions and talk about goats knowledgeably:

- **Doe:** A female goat. A young doe is called a *doeling*.

  You may have heard a female goat called a *nanny*. But don't use that term unless you want to offend someone. Some meat goat owners still call their does nannies, but the more common term these days is *doe,* especially if you're getting dairy goats.

- **Buck:** A male goat. A young buck is called a *buckling*.

  Like nanny goat, *billy* is also considered a negative term for a male goat, bringing to mind a scruffy, stinking animal. Some meat goat owners still use this term, but in order not to offend, use *buck* instead.
- **Brood doe:** A doe that is kept and used for breeding purposes, to pass on certain desirable genetic traits.
- **Kid:** A baby goat or a goat less than a year old.
- **Yearling:** A goat that is between one and two years old.
- **Wether:** A castrated male goat.
- **Herd:** A group of goats. Sheep are in flocks; goats are in herds.
- **Ruminant:** An animal that has a stomach with four compartments and chews cud as part of the digestive process. Goats are ruminants.
- **Udder:** The organ in a goat that produces milk. Don't call it a *bag*. Goats have only one udder.
- **Teat:** The protuberance from the udder that you use to milk a goat. Goats have two teats.
- **Dam:** A goat's mother.
- **Sire:** A goat's father. You can also say that a goat *sired* a kid.

Taking a Look at Goat Anatomy

Goats are mammals and are similar to other mammals in some ways. But they also have unique features that indicate whether they are healthy, tell you how old they are, and even give clues about their parents.

In this section, I tell you about different parts of the goat, how to tell a goat's age by his teeth, and how to tell a goat from a sheep. (Some of them do look similar.)

Parts of the body

You can own goats and not know the names of parts of the body. But if you want to have an intelligent discussion with other goat aficionados or show your goats, knowing the correct terminology is essential.

Figure 2-1 shows you the names of the different parts of a goat. Some are obvious — we all can identify an ear or a neck. But others may be new to you if you haven't raised animals before. Here are some terms for different body parts:

- **Cannon bone:** The shin bone.
- **Chine:** The area of the spine directly behind the withers.
- **Escutcheon:** The area between the back legs, where the udder lies in a doe. This area should be wide in a dairy goat.
- **Pastern:** The flexible part of the lower leg below the dewclaw and above the hoof.
- **Pinbone:** The hip bone.
- **Stifle joint:** In the back leg, the equivalent of the knee in a goat.
- **Thurl:** The hip joint, usually referred to in relation to the levelness between the thurls.
- **Withers:** The shoulder area or area of the spine where the shoulder blades meet at the base of the neck. You measure from this point to the ground to determine a goat's height.

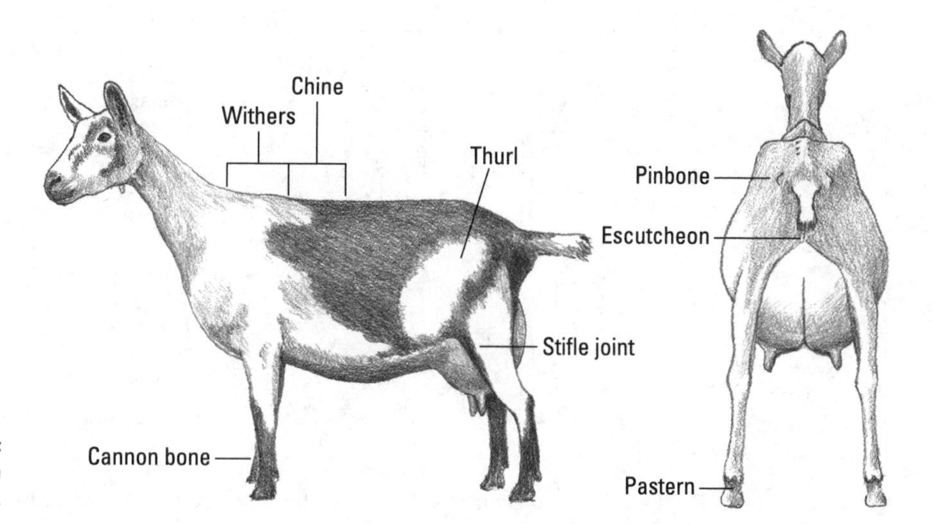

FIGURE 2-1: The parts of a goat's body.

The digestive system

Goats are *ruminants,* which means that they have four stomach compartments and part of their digestive process includes regurgitating partially digested food and chewing it, called *ruminating.* This kind of digestive system needs a plant-based diet.

Understanding a goat's digestive system and how it works helps you keep your herd healthy or identify potential problems.

TECHNICAL STUFF

The goat stomach consists of three *forestomachs* — the rumen, reticulum, and omasum — and a true stomach, the *abomasum.* (See Figure 2-2.) The forestomachs are responsible for grinding and digesting hay, with the help of bacteria. The last compartment, the abomasum, is similar to the human stomach and digests most proteins, fats, and carbohydrates.

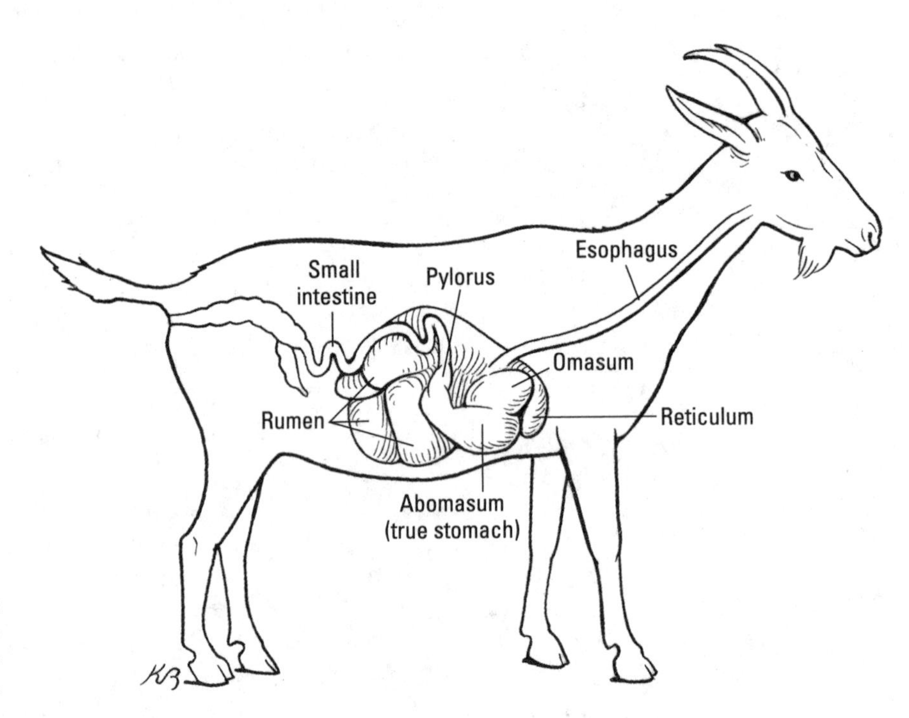

FIGURE 2-2: The parts of a goat's digestive system.

Each stomach compartment has a different function, and they all work together:

- The **rumen** is the largest of the forestomachs, with a 1- to 2-gallon capacity. It is a large fermentation vat that has bacteria living in it. These microorganisms break down roughage, such as hay. Then the goat regurgitates the partially broken-down material, chews it as a cud, and then swallows it.

This repetitive process, rumination, creates methane gas as a byproduct. Methane is the cause of the strong-smelling belches that you can expect from a goat with a healthy rumen. A goat that can't belch has bloat. (See Chapter 11 for signs and treatment of bloat.) The rumen action also creates heat, much like a compost pile, which helps a goat stay warm.

- The **reticulum** is in front of and below the rumen, near the liver; the reticulum and the rumen work together to initially break down the food. Rumen contractions push the smaller particles of partially digested food into the rumen and heavier pieces into the reticulum. Then the reticulum contracts and sends the partially digested food into the mouth as a *cud* for chewing.

  This process continues until the pieces are small enough to pass through to the omasum. The reticulum also catches harmful things, such as wire or nails, that a goat accidentally swallows (see Chapter 11).
- After fermentation and rumination break down the roughage, it moves through the reticulum to the **omasum,** where enzymes further digest it. The omasum has long tissue folds whose function is to help remove fluid and decrease the size of food particles that come out of the rumen.
- The **abomasum** is the only compartment that produces digestive enzymes. It completes the next step in the digestive process of food that forestomachs partially broke down. The abomasum handles the primary digestion of grain and milk, which don't need rumen bacteria to be digested. The products of this part of digestion pass into the intestine for final breakdown, separating waste products from usable fats and proteins.

Hooves

A hoof is the horny sheath that covers the lower part of a goat's foot and is divided into two parts. Goats stand on their hooves and walk on them to get around, which makes them extremely important.

WARNING

When something goes wrong with a hoof, the rest of the musculoskeletal system is affected, which can cause pain, limping, lameness, and a shorter lifespan.

Because they don't like to stand in one place to eat, goats don't do well if they have to lie around or walk on their knees. (Yes, knees.) I recently saw goats on television whose owners had fed them well but apparently hadn't trimmed their hooves for years. (Chapter 9 tells you how to trim hooves.) The goats' hooves were almost a foot long and curled up at the ends. Consequently, many of them couldn't even walk, and others walked on their knees, dragging their rear feet. That had to hurt!

Untrimmed hooves make a goat prone to foot scald or rot, which can ultimately kill a goat. (See Chapter 11 for more on foot rot.) Goats' hooves do best in dry, rocky climates. Goats that are feral or were feral for years, such as the Spanish goat, need less hoof care than closely bred, farmed goats.

TIP

A proper goat hoof is rhomboid-shaped (not rectangular, but slightly longer in the front than in the back) and has no overgrowth on the sides or front. Trim your goats hooves regularly to ensure that they maintain this shape.

Teeth

Goats have lower teeth in the front of their mouths but only a hard pad on the top. They also have back teeth on both top and bottom, which you will painfully discover if you put your finger into the back of a goat's mouth! The back teeth are used for cud-chewing.

Baby goats get their first teeth before birth, at around 98 to 105 days of gestation. They lose these baby teeth, just like other mammals. You can generally determine the age of a goat by looking at the eight teeth in the goat's lower front jaw. This is called *toothing* a goat.

Toothing a goat is not completely accurate because you can find variances among goats. They may lose some teeth and grow new ones at different times, or their diet or health can influence how their teeth grow. Figure 2-3 shows you goats' teeth at various stages.

Here's a guide to toothing a goat:

- **First year (kid):** Baby teeth are small and sharp. They gradually fall out and permanent teeth replace them.
- **Second year (yearling):** The two middle front teeth fall out when the kid is about 12 months old. Two larger permanent teeth grow in their place.
- **Third year:** The teeth next to the two middle teeth fall out. Two new, larger, permanent teeth grow in when the kid is about 24 months old.
- **Fourth year:** The next two teeth on either side of the four middle teeth fall out, and new permanent teeth grow in.
- **Fifth year:** The goat has all eight front teeth.

  After five years, you can guess at the goat's age by looking for wear on the teeth and missing teeth. This will vary a lot, depending on the goat's diet.

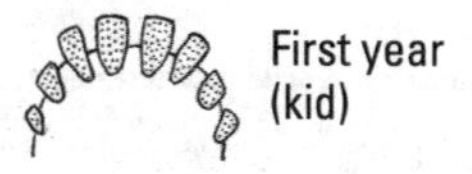

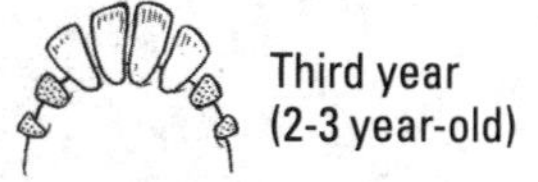

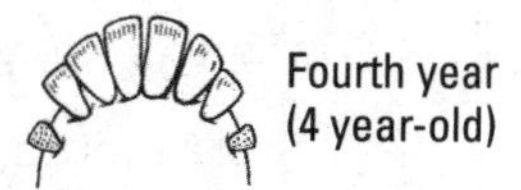

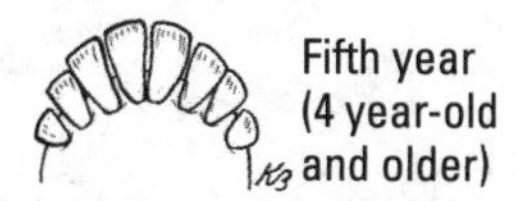

FIGURE 2-3: You can tell a goat's age by his teeth.

Beards

Most but not all goats have beards. The male's beard is more majestic than the female's beard. Although the purpose of the beard is unknown, it's great for capturing that unique scent that males like to flaunt during breeding season. If you're having trouble telling whether an animal is a goat or a sheep, look for a beard — only goats have them. (But not all goats have them.) Table 2-1 tells you more about distinguishing goats from sheep.

Wattles

A *wattle* is a fleshy decoration that hangs from the goat's neck just past where the chin attaches. Wattles are more common in dairy goats and pygmy goats. Most goats have two, although I have had a goat born with only one. They are a genetic trait — at least one parent has to have wattles for a kid to be born with them.

TABLE 2-1 Distinguishing Goats from Sheep

Goat	Sheep
Tail stands up	Tail hangs down
Horns usually straight	Horns usually curled
Often bearded	Non-bearded
Browser	Grazer
Curious	Aloof
Independent	Blindly follow
Hate to get wet	Don't mind rain
Butt heads by rearing up	Butt heads by charging
No division in upper lip	Divided upper lip

Horns

Most goats will grow two horns (yes, even the females) unless they're disbudded shortly after birth. (I tell you about disbudding in Chapter 9.) A minority of goats are naturally hornless, a trait called *polled.* Although polled goats are desirable because they save some work for the owner and some pain for the goat, in some dairy breeds, if you breed polled goats to each other, you may have a higher incidence of *intersex goats* (goats that have male and female sex traits).

Intersexuality in goats is a recessive characteristic seen only in females, which causes infertility. The one intersex doe who came from my farm acted like the worst kind of buck — constantly blubbering and mounting her pen mates.

The way to tell if a goat is polled (at least one parent must be polled for a goat to inherit the trait) is to look for the lack of a swirl on the head where the horns grow. Eventually, the polls will grow into rounded bumps — like a giraffe has, only much shorter.

Eyes

A goat's pupil is rectangular rather than round like other animals'. A lot of people say they are creeped out by goat's eyes, but according to the Los Angeles zoo, they have an important function. Goats have excellent night vision, which enables them to avoid predators and browse at night. I can vouch for the night browsing, although my goats are out at night eating under a full moon more often than any other time.

Goats' eye color ranges from yellow to brown to blue. Angora goats sometimes inherit a condition where facial hair covers their eyes, which harms their vision and their ability to browse.

Signs of a Healthy Goat

One of the most important parts of being a goat owner is observing the goats to make sure that they're healthy. You can do it when you're feeding, or just go out and watch them. The bonus is that hanging out with goats is relaxing! I'm lucky to work from home and be able to incorporate "goat breaks" into my routine.

A healthy goat has shiny eyes and glossy hair, and is curious and energetic unless resting and chewing cud. If you're watching your goats and one of them seems a bit off, you can take a few simple steps to investigate further. In this section, I tell you what's normal for goats and give you some simple clues to determine whether your goat is healthy.

Noticing posture

One of the first signs of a healthy goat is posture. A healthy goat usually has her head and tail up, stands erect, and holds her ears erect. That doesn't mean that every time a goat has her tail or ears down that she is sick. It is just a sign to be taken in conjunction with other signs. A goat that doesn't feel well will hunch with tail down and not be as responsive to external stimuli such as sounds or physical activity.

TIP

A goat with an upset stomach, bloat, or urinary calculi (see Chapter 11) will stretch out repeatedly, trying to relieve the pressure or discomfort or trying to pee. This abnormal posture is a sign that you need to check out the goat immediately.

Observe your goats when they're healthy and note how they stand and move, how they interact, and how they look overall. Any change is your first clue that something might be wrong.

Identifying the meaning of goat cries

People with new goats often ask whether something is wrong with goats because they cry every time they see a person move. I advise them not to respond to the goats every time they cry. Often they were just spoiled by their prior owner or are a breed known for being noisy, such as Nubian or Nigerian Dwarf. (Chapter 3 tells you about goat breeds.) Or they are grieving the loss of their mother or other herd members.

I had a goat whose baby died during birth and had to be removed from her body. She saw the dead kid at the vet's office, but after she got home she ran to the kidding pen, looking around and then yelling at me. She recovered physically from the ordeal but took months to get over it emotionally. Some goats whose babies are taken at birth for bottle-feeding (usually dairy goats) or die later respond the same way, while others seem unconcerned.

If a goat is truly hungry or thirsty, his bleat is persistent. I always hear these cries as my wake-up call in the morning if I'm not out by 7:00 a.m. and in the evening at the same time. They *know* they're due to be fed. Who needs a traditional alarm clock?

A sick goat sometimes moans or makes a stressed-out sounding cry, but more often you notice that she *isn't* crying but is away from the herd suffering silently.

Bucks in rut will make some of the craziest snorting, bleating noises you've ever heard. Some does cry out in little short bursts when they're in heat and wanting to get bred.

Does that are kidding can also be quite noisy, although some approach the task silently. During the first stage of labor, they whine more than cry, especially if they want you there with them the whole time. Others are pretty discreet until it's time to push the baby out and then they let loose with a loud, long cry to tell you over the baby monitor that it's time.

Finally, a goat that is trapped or injured (or got through the fence and can't get back in) will make sure you know it with loud, frantic bleats. This is the kind of cry that sends you running to its source because time is of the essence. If a kid is the victim, the mother will often chime in to let you know that something is wrong.

Listen, and learn your goats' cries; the knowledge will serve you well.

Determining normal temperature

A goat's normal temperature is around 102°F to 103°F but can be a degree higher or lower, depending on the individual goat. A goat's temperature can also go up or down throughout the day. On a hot day, you can expect some of your goats to have higher temperatures.

A temperature that is too high or too low indicates illness in a goat. A goat with a high temperature often has an infection and can quickly become dehydrated, while a goat with a low temperature (hypothermia) may have rumen trouble or is so sick that he is unable to stay warm. This goat needs to be warmed or he will die.

TIP

If you have a goat with hypothermia, use a *goat coat* — a jacket made to fit a goat and keep him warm, especially when he's sick. You can buy one from a goat catalog or individual seller online, or you can make one out of a fleece blanket or an old sweatshirt. Just cut the blanket so it is big enough to wrap around the goat's body, cut two front leg holes, and put it on with the open part on the goat's stomach. If you're using a sweatshirt, cut through the front of the sweatshirt to create an opening at the bottom.

Figuring out what's normal for your goats

In order to determine what's normal for each of your goats, take their temperature several times when they're healthy and note the number in their health records. Make sure you measure their temperatures on a hot day and a normal day so that you get an accurate baseline to compare with if a goat gets sick, as well as an idea of what variations might occur in that goat.

Taking a goat's temperature

Taking a goat's temperature is easy. You need either a digital or traditional glass thermometer that you can buy from a feed store, a drugstore, or a livestock supply catalog. Either type is fairly inexpensive.

WARNING

If you use a glass thermometer, make sure you shake it down before you start so that it reads accurately. Tie a string around one end of a glass thermometer so that you can retrieve it if it goes too far.

To take a goat's temperature, grab a thermometer, the goat's health record, and take the following steps:

1. **Immobilize the goat.**

   You can hold a small kid across your lap. Secure an adult in a *stanchion* (a metal device to lock a goat's head in place), have a helper hold him still, or tie him to a gate or fence.

2. **Lubricate your thermometer with K-Y Jelly or Vaseline.**
3. **Insert the thermometer a few inches into the goat's rectum.**
4. **Hold the thermometer in place for at least two minutes.**
5. **Slowly remove the thermometer, read the temperature, and record it on the goat's health record.**
6. **Clean the thermometer.**

   Use an alcohol wipe or a cotton ball that has been wet with alcohol.

Using ruminations as a health indicator

Because rumination is an essential part of how goats digest food (see the section "The digestive system"), you can use cud-chewing as an indicator of goat health. A ruminating goat is eating and generating heat and energy. You can determine whether a goat is ruminating in two ways: by looking for cud-chewing and by listening to the goat's body.

A goat's rumen is located on the left side of the abdomen. You can watch this area or feel the side of the abdomen for movement.

The best way to determine whether a goat is ruminating and the strength and frequency of rumination is to listen. Often, ruminations are loud enough that you can hear them by just sitting next to the goat. If you can't hear them, put your head up to the left side of your goat's abdomen. If you still have trouble hearing ruminations, use a stethoscope. You can purchase an inexpensive stethoscope from a livestock supply catalog.

REMEMBER

Healthy ruminations are loud, sound kind of like a growling stomach, and occur about two or three times a minute. If they are weak or infrequent, give your goat some roughage and *probiotics* ("good" microbes given orally that protect against disease) to stimulate the rumen and to add to the rumen bacteria.

Look around your herd to see whether each goat is chewing her cud. I've found that a good time for this is the early afternoon, when the goats are resting before their last go at the pasture for the day. Usually at least two-thirds of them will be ruminating at the same time. Take a closer look at any goats that aren't chewing cud. If they don't look well in some other way (see the section, "Signs of a Healthy Goat"), go up to them and listen for rumination sounds.

Taking a goat's pulse

Pulse indicates the goat's heart rate. Normal pulse for a goat is 70 to 90 beats per minute. Kids' heart rates may be twice that fast. To take your goat's pulse:

1. **Make sure she is calm and resting.**
2. **Find the goat's artery below and slightly inside the jaw with your fingers.**
3. **Watching a clock, count the number of beats in 15 seconds.**
4. **Multiply that number by four to get the pulse rate.**

Counting respirations

The normal respiration rate for an adult goat is 10 to 30 breaths per minute and for a kid it is 20 to 40 breaths per minute. To count respirations, simply watch the goat's side when she is calm and resting. Count one respiration for each time the goat's side rises and falls.

Recognizing life expectancy

Goats spend the first three years of their lives growing. They are almost their full size at age 2, but can still put on weight and height for another year.

Goats' general life expectancy is about 7 to 12 years. Wethers often live longer than does and bucks — over 10 years — probably because they are most often pets and have less stressful lives. One of the common causes of death in wethers is urinary calculi.

Does live an average of 11 to 12 years. They often die from kidding-related problems because of the stress that it puts them under. They are more prone to parasite overload and losing condition from feeding kids. Does who aren't bred after about the age of 10 often live longer. I have heard of does living as long as 20 years.

Bucks have shorter lives than either wethers or does, living on average 8 to 10 years. The stresses of *annual rut* (breeding season; see Chapter 12) take their toll on them. They have to deal with other bucks' aggressions and lack of eating and sleeping.

Elderly goats require special care and feeding to deal with health conditions that they may have. I tell you more about older goats in Chapter 14.

Using a Goat Scorecard to Evaluate a Goat

Goat registries use scorecards to evaluate goats that are being shown. These vary with the type of goat being evaluated. For example, out of 120 points in the Colored Angora goat scorecard, 70 points are for fleece. In the dairy goat scorecard, more focus is put on the mammary system and in meat goats, muscle development and growth have more value.

Look on the Internet or write to the registry for the breed you are interested in and request a copy of their goat scorecard online and use it to help you evaluate and learn about the ideal for that type of goat. The Colored Angora Breeders Association (CABGA) includes a sample of a judge's scorecard on its website (`www.cagba.org/scorecard.pdf`).

» Learning the pros and cons of different breeds

» Figuring out which breed you want to raise

Chapter 3

Knowing Your Capra Aegagrus Hircus (Goat, That Is)

So you want to get goats? You've probably already thought about why you want goats. So now you need to figure out what kind of goat can best serve your purposes.

If you want a goat to eat brush, then your choices are unlimited; you just have to figure out what size is best and how many you want. And if you know you want goats for just one purpose — milk, meat, or fiber — then the decision will be easier and you can just read about those kinds of goats. But what if you want a goat for milk and meat? Or fiber and milk? That might make the decision a little harder.

TIP

The American Goat Federation represents all goat producers and provides information on different breed clubs and registries (`https://americangoatfederation.org/resources`).

I give you some pointers below as I describe the three types of goats — dairy, meat, and fiber — and the breeds within those types.

Looking into Dairy Goats

If you want dairy goats, you have a variety of breeds and sizes to choose from. The American Dairy Goat Association (ADGA), one of the United States' two major dairy goat *registries* (organizations that keep official lists of goats within a specific breed, provide registration certificates, and compile pedigrees), recognizes eight different dairy goat breeds: Alpine, LaMancha, Nigerian Dwarf, Nubian, Oberhasli, Saanen, Sable, and Toggenburg. The other registry, the American Goat Society (AGS), registers only purebreds and also recognizes the Pygmy as a dairy breed. (You can read more about registries and whether to buy registered or unregistered goats in Chapter 7.)

Neither of these registries recognizes the mini crossbreeds that have captured the hearts of urban goat owners, so two new registries have sprung up: The Miniature Goat Registry (TMGR) and the Miniature Dairy Goat Association (MDGA).

In addition, the Canadian Goat Society (CGS) registers purebred dairy goats, which also are accepted by the major dairy goat registries noted earlier.

Dairy goats provide milk, of course, and if you're interested in a home supply of goat meat, your dairy goats can serve that purpose, too. Unfortunately, if you want fiber and milk, most dairy goats won't work because their coats generally are too short, but fiber goats usually don't produce a lot of milk.

In the upcoming sections, I give you an idea of what each of the standard dairy goats looks like and some of their traits. If you decide that you just have to have one of these breeds but can't find it in your area, see Chapter 7, which tells you about sources for finding goats.

Standard dairy goat breeds

The standard breeds are a good choice if you want dairy goats that produce a lot of milk and you have the space: ADGA and AGS require that each of the breeds reaches a minimum, but not a maximum, height and weight. The minimum height for does ranges from 28 inches to 32 inches, depending on breed. The minimum buck heights are 30 to 34 inches.

The standard breeds produce an average of 5 to 8 pounds of milk each day for 10 months. This is the equivalent of approximately 2½ quarts to a gallon per day. Dairy goats' milk production starts out lower, gradually increases throughout lactation, and then goes down again.

If you think you want standard dairy goats, deciding which one is for you depends mainly on size, the look and personality that strikes your fancy, and what is available to you. The upcoming sections tell you about the standard dairy goat breeds and some of the characteristics that define them.

Alpine

Alpines, also known as French Alpines, hail from the Alps. They are large goats that have erect ears and come in a variety of colors and patterns. They're friendly and hardy, and the wethers (castrated males) are a favorite for use as pack animals. Many goat milk dairies have Alpines in their milking string because they consistently produce a lot of milk and often can be milked through without rebreeding. Figure 3-1 shows an Alpine goat.

FIGURE 3-1: An Alpine goat.

LaMancha

LaManchas look earless, but they actually have very small ears. The ears are of two types: gopher ears and elf ears. (Figure 3-2 shows a LaMancha with gopher ears.) Gopher ears are very small and rounded; elf ears are less than two inches long and turn up or down. People tend to either love or hate the looks of the LaMancha.

FIGURE 3-2: A LaMancha goat with gopher ears.

LaManchas descended from the Spanish Murciana and Spanish LaMancha goats brought to the United States from Mexico for meat and milk. They first got recognition as a separate breed in the early 1950s and are fast becoming a very popular breed. They come in a variety of colors and patterns and are very friendly. I have also found them quite curious, with an ability to get to food regardless of where you put it!

Nubian

Nubians (see Figure 3-3) are one of the most popular breeds. They have long, floppy ears and a rounded (Roman) nose, and they have been bred to be very large. Their milk has a high butterfat content, and they produce a lot of it — which makes them a good choice if you want to make cheese. Because of their large size, they make good meat goats, too.

Unfortunately, Nubians also are known for the loud, annoying cries that many of them make. The cries make them undesirable in some neighborhoods and on some farms.

FIGURE 3-3: A Nubian goat.

G-6-S GENE MUTATION: A PERSISTENT DEFECT IN NUBIANS

The G-6-S gene mutation was first discovered in 1987. Further investigation showed that about 25 percent of Nubians carry the gene, which results from a single mutation. It affects only Nubians and Nubian crosses. Carriers (N/G genotype) aren't affected but can transmit the gene to 50 percent of their offspring. Carriers can make great pets, packers, and brush control goats.

Goats with the gene mutation lack the enzyme G-6-S, which causes a variety of symptoms that vary in severity. Failure to grow is the main sign. In some cases, a kid starts out smaller than normal and continues to grow slowly; in others a kid grows normally for three months or so and then stops growing. Other goats seem to grow to a normal size but are found to actually be small in comparison to other goats in that bloodline. Other signs include a lack of muscle mass, a "slab-sided" appearance or a "blocky" head. They seem to have compromised immune systems, experience reproductive problems and, in some cases, become deaf or blind. The longest a goat with G-6-S has been known to survive is less than four years. Death is usually caused by heart failure. Because goats with this defect can live long enough to breed, they continue to pass on the gene. Many Nubian breeders now test all breeding stock for the defect.

Oberhasli

Oberhaslis were originally known as Swiss Alpines, a variation on the Alpine. They were eventually recognized as a separate breed with distinct markings. They have erect ears, are medium-sized, and have a reddish-brown color (chamoisee) with black markings on their backs, belly, tail, and legs. Does can also be pure black.

Oberhaslis are my favorite of the large breeds because of their sweet temperament and beautiful markings. They have good dairy character, which means they seem to like being milked.

Saanen and Sable

Saanens are white or off-white in coloring and are the largest of the standard dairy goats. Sables are Saanens that aren't white, due to a recessive gene. They have erect ears, like the Alpine, and are usually mellow and easy-going.

Sables and Saanens are known for their high milk production and are therefore often used in commercial dairies. A downside to Saanens is that their white coats show dirt and also make them more prone to sunburn.

Toggenburg

Toggenburgs, also referred to as Toggs, are beautifully marked and range in color from fawn to chocolate brown with white markings. They resemble a medium-sized Alpine in body shape, with erect ears. The Toggenburg is a Swiss dairy goat from Toggenburg Valley of Switzerland at Obertoggenburg. They are also credited as being the oldest known dairy goat breed.

Toggs produce a moderate amount of milk but are known for long lactations. More than one person has told me that they don't like the taste of Toggenburgs' milk, probably because it has a low butterfat content and doesn't taste as rich as some of the other goats' milk.

Miniature breeds

The standard dairy goat breeds are impressive, but maybe you don't want to be drowning in milk. With the miniature breeds, you get some of the same characteristics but in a smaller package.

Miniature breeds are relatively new to the scene but are growing in popularity. Miniature dairy goats include Nigerian Dwarves, Pygmies, and the most recent development in dairy goats in the United States: miniature, crossbred versions of the standard breeds, called *minis*.

LITTLE GOAT, A LOT OF CHEESE

My first goats were Nigerians. Their milk was high in protein and extremely high in butterfat — normally 5 to 6 percent and as high as 10 percent at the end of their lactation (normally 10 months). This higher butterfat milk produces more cheese than milk with low butterfat.

I did a comparison between the milk of a Toggenburg and a Nigerian and found that I got twice as much cheese from the Nigerian milk!

At least one goat cheese dairy in the United States uses Nigerians exclusively because of the quantity of cheese they can produce. You have to milk more goats to get the same quantity of milk you would from standard dairy goats, but you can get a lot more cheese from each gallon of milk.

Nigerian Dwarf

Nigerian Dwarves are miniature goats that came to the United States from Africa. They're considered dwarves because they're a product of hereditary pituitary hypoplasia, which means they're small but normally proportioned. They are colorful, friendly, easy to handle, and in some cases quite good milk producers. Figure 3-4 shows a Nigerian Dwarf goat.

FIGURE 3-4: Nigerian Dwarf goat.

Two height standards exist for Nigerian Dwarves. The Nigerian Dwarf Goat Association allows 17 to 21 inches for does and 19 to 23 inches for bucks, while the American Goat Society allows a maximum of 22.5 inches for does and 23.5 inches for bucks. Their average weight is around 75 pounds.

Their kids are the cutest things around, and Nigerian Dwarves have a lot of them. Quadruplets aren't rare, and you even see quintuplets and sextuplets from time to time. The does tend to be easy kidders (I'm not kidding!) and natural mothers.

Nigerians are popular as milkers, 4-H projects, and pets. The wethers are easier to sell because they make good backyard pets and are easy for children to handle. They can be expensive, but if you find a breeder who will sell them unregistered you can get them for a reasonable price.

TIP

Some Nigerian Dwarves give only about a pound (one pint) of milk per day, but others produce as much as 8 pounds (a half-gallon). A lot of them are hard to milk because of their small teats, but others have been bred to alleviate this problem. If you want these goats for milking, make sure to ask about milking, look at the udders, and try milking some goats, if you can.

Some urban goat farmers shy away from the Nigerians, despite their petite size, because they can be loud. But their flashy colors and winning personalities are a draw.

You can eat a Nigerian, but most people don't. The cost isn't worth the end result, and besides, they're so darn cute!

Pygmy

Pygmies, like Nigerians, are miniature goats that came to the United States from Africa. Pygmies have been bred to be short and stocky (unlike Nigerians) and are more limited in color, ranging from white to brown to black with only minor variations. The most common pattern is *agouti,* which refers to alternating bands of colors, kind of like salt and pepper hair. (Figure 3-5 shows an agouti Pygmy goat.)

Pygmy goats aren't usually considered dairy goats, but because they don't fit neatly into any category I include them here. Most of them are raised as pets or 4-H projects, but a small minority of breeders raise them for milk. The Pygmies raised for milk tend to look more like Nigerians in body shape. If you want Pygmies for milk, make sure to ask whether they're registered with the American Goat Society and whether the breeder is milking them.

FIGURE 3-5: An agouti Pygmy goat.

Because they have been bred down to a short, compact size, Pygmies often have problems kidding and need the services of a veterinarian for cesareans. They are not quite as prolific as the Nigerians, either.

Like Nigerians, Pygmies are much less expensive to keep and take up a lot less space than a standard dairy goat, so they work well in rural and urban settings.

Kinder

The Kinder is a moderate-sized breed that was developed in Washington and first registered in 1988. It is a cross between the Pygmy and Nubian breeds. Kinders are good dual-purpose (milk and meat) goats that are ideal for family farms.

Minis

If you really like one of the standard breeds but are in the city or just don't have the space for big goats, minis are ideal. These goats are a cross between a Nigerian Dwarf buck and a doe of the breed you want to miniaturize.

WARNING

Don't breed a full-sized buck to a Nigerian Dwarf doe. The doe may have problems kidding as a result of too-large kids.

UNDERSTANDING MINI GENERATIONS

The kids produced by the first cross of a Nigerian buck with a standard dairy goat is the first generation (F1). Breeding an F1 to another F1 gives you second generation (F2) kids. Breeding F2 to F2 gives you F3 kids and so on, for six generations.

A kid's generation is always one generation higher than the lowest generation parent. So if you breed an F2 to an F1, the kids will be F2. Even if you breed an F4 to a F1, the kids will still be F2.

F1 and F2 goats are considered experimental. F3 through F5 goats are called "American" and an F6 is a purebred. Ask about the generation when buying a registered mini dairy goat. Purebreds are worth more than first generations.

The minis don't have quite as many multiple kids as the Nigerians — which can even have septuplets -- but they are more likely to have triplets and quadruplets than the standard breeds.

You can find Mini Nubians and Mini Manchas pretty easily. Finding Oberians, Mini Toggs, Mini Alpines, and Mini Saanens can be more challenging because there are fewer of them. (Chapter 16 tells you more about breeding miniature dairy goats.)

If you plan to milk and you live in a city that allows backyard goats, these little goats are a good choice. Jenny Grant of the Goat Justice League, which got the Seattle City Council to allow miniature goats within the city limits, recommends the Oberian and the Mini Mancha for urban situations because they're the quietest.

Discovering Meat Goats

A lot of people around the world eat goat meat (about 75 percent, at last count), and it is catching on in the United States, as well. The meat is very lean and delicious. Meat goats are generally not used for milking, and some even have traits that make them undesirable to milk. Most of the breeds don't make good fiber goats because they have short coats, but the Boer is sometimes and Spanish goat is always an exception.

Meat goats grow fast and are fantastic brush-eaters. Some breeds are nearly self-sufficient because they evolved in feral conditions. They require much less time for management because they have developed resistance to the parasites, foot rot, and respiratory problems that are so common in other goats.

If you're interested in goats to sell in the fast-growing U.S. market, to get rid of kudzu or blackberries, or just for your own freezer, you have a growing number of breeds to choose from.

Boer goats

Boer goats are large, heavy animals that resemble Nubians because of their Roman noses and floppy ears. They're white and reddish-brown (although some are all red) or black, as well as spotted and dappled, and many are horned. Their horns are short and curve back close to the head. Bucks can weigh from 260 to 380 pounds and does from 210 to 265 pounds. Figure 3-6 shows a Boer buck.

FIGURE 3-6: A South African Boer buck.

Boers are originally from South Africa, where they were bred for hardiness. They came into the United States from New Zealand and Australia through Canada and continue to grow in popularity. Based on the increased number of men I saw at our local goat conference, the Boers are catching on. (Past conferences have drawn mainly dairy goat owners, who traditionally are women.)

If you want a purebred, high-quality Boer, expect to pay a lot. Boers are one of the most expensive breeds in the United States. That doesn't mean that all of them are expensive, though; you can still get some good *foundation stock* (the stock you start your breeding program with) if you look around.

At least two organizations register Boer goats, including the American Boer Goat Association (ABGA) and the United States Boer Goat Association (USBGA).

Boers are quite adaptable and hardy, and most are affectionate and mild-mannered. Their adaptability leads them to browse more than dairy goats because they can easily handle both heat and cold. Boers do have some fairly common genetic defects you need to watch for, including extra teats and abnormal testicles.

TIP

Some breeders of large dairy goats, such as Nubians, cross their does to a Boer bucks. This has the added advantage of selling the kids for milk while freshening the does for another year of milking.

Myotonic goats

Myotonic goats, also known as *fainting* or *stiff-legged goats,* are so named because of their tendency to go rigid and fall down when they're startled. This is a genetic defect in a recessive gene that probably started in just one goat and then was continued through breeding. The repeated muscle tightening means that the goats have more muscle and are therefore good meat goats.

Fainting goats are often much smaller than Boers, weighing in at between 50 and 75 pounds and becoming no more than 25 inches tall. Those raised for meat, such as the Tennessee Meat Goat, can be much bigger — weighing up to 175 pounds or more.

They come in an array of colors and patterns and have long ears that stick out sideways. They often have long horns that curve backward. They can be shy animals (wouldn't you be if you kept falling down at the most inopportune times?) but have sweet personalities.

These goats can be registered by either the Myotonic Goat Registry (MGR) or the International Fainting Goat Association (IFGA). These goats are currently listed as "recovering" on the Livestock Conservancy's list of rare livestock breeds.

TIP

Myotonic goats, because of their diminutive size, may be a better choice for a meat goat breeder who has limited space to work with. The bucks can also be crossed with Boer does to grade up to higher meat production.

Some people buy fainting goats for the novelty of scaring them and watching them drop. The goats have allegedly served another purpose — protecting sheep when a predator comes around. The goat drops and the predator eats the goat instead of the sheep. That's literally a sacrificial goat!

Kiko goats

The Kiko goat is a newer breed that was created in New Zealand purely as a meat goat. Kikos are most often white with long, scimitar-like horns and medium ears that stick out sideways. They can gain substantial weight without supplemental feeding, which is a big plus for producers. If you have substantial range for these goats, they may be even more economical than Boers.

The name *Kiko* means "meat for consumption" in the Maori language. This unique breed came out of a government-funded project intended to thin out the native goat population. Once captured, some of the native goats were crossbred with milk and fiber goats with the intent of developing hardy goats that grew fast and produced a lot of meat.

Kikos came into the United States in the 1990s and were found to be superior to the Boer goat in terms of hardiness and adaptability. Their hooves have to be trimmed less frequently than other goats, they kid easily, and the does are good mothers. These traits make them perfect for a life on the range.

Kiko goats and offspring from crosses with other meat goats can be registered with the International Kiko Goat Association. They officially recognize a Boer/Kiko cross called the *American Meatmaker.* The American Kiko Goat Association (AKGA) allows some crossbreed registration, but the sire must always be a purebred Kiko.

Spanish goats

Spanish goats, also called *brush goats, scrub goats,* and *woods goats,* are usually medium-sized and lanky with long horns that often twist at the end. They come in all colors and mostly have short hair but can also have long hair. They are known to be crossed with Nubian, Angora, Boer, and Alpine or other Swiss breeds, which you can often recognize by their looks. Those that are crossed with Cashmere goats can also serve as fiber goats. Figure 3-7 shows a Cashmere Spanish goat.

If you decide to get a Spanish goat, you will also be getting a little bit of U.S. history. These goats were originally brought over from Spain in the 16th century and evolved into nearly indigenous goats in the southern United States. Because they were feral, they are much hardier than goats such as the Boer so they make good, easy-care meat goats, much like the Kiko.

The American Livestock Breeds Conservancy has developed standards for the Spanish goat and formed the Spanish Goat Association (SGA). Its goal is to find Spanish goats and breeders in the United States to preserve the gene pool. The number of breeders is limited, and so far none operate in the West or Northeast.

FIGURE 3-7: A long-haired Spanish goat, also considered Cashmere.

TIP

If you want to learn more about true Spanish goats or even get a few, check out the SGA at `www.spanishgoats.org`.

Texmaster goats

The Texmaster is a moderate-sized meat goat and a trademarked cross between Boers and Tennessee meat goats (myotonics) developed by Onion Creek Ranch in Texas. They usually don't "faint," but they do add to the meatiness of the breed.

Moneymaker meat goats

Moneymaker meat goats were developed by Bob and Dusty Copeland of California by first crossbreeding Saanens and Nubians and then adding Boers into the mix. This crossbreed has the advantage of easy kidding, often with triplets. In addition, the does are able to put the desired weight on kids within four or five months of nursing. The breeders have found that the ideal mix is 75 percent Boer and 25 percent Saanen/Nubian cross.

Savanna goats

Another South African meat breed brought to the United States is the Savanna. These goats are known for their heat and drought resistance and excellent mothering abilities. Savannas are a muscular breed with long ears, thick black skin, and a short white coat that develops a nice, fluffy cashmere undercoat for additional warmth during the winter. They are highly adaptable to a variety of unpleasant weather conditions, having done well in both Canada and the southeastern United States.

Because international law now prohibits bringing in any new genetic material, these goats are expensive and harder to come by.

Investigating Fiber Goats

If you are a home spinner or want to get your own supply of fiber, consider the fiber goats. Angoras produce the fiber called mohair, which is a silky fiber used in many products. Cashmere, produced by the cashmere goat, is an even more exotic fiber and is in high demand. It comes from the undercoat of these goats.

Fiber goats, if you properly care for them, take a bit more work than some of the other goats. You can choose from medium to small goats in a variety of colors. If you want meat, some of these are dual-purpose goats.

A number of registries exist for fiber goats, mostly depending on the breed, but in two cases specifically for colored fiber goats. These include the American Angora Goat Breeders Association (AAGBA), the Cashmere Goat Association (CGA), the Pygora Breeders Association (PBA), the PCA Goat Registry, and the Colored Angora Goat Breeders Association (CAGBA).

While CAGBA is only for Angoras, the PCA Goat Registry is hybrid-based, registering goats whose parents are already registered by CAGBA, PBA, NPGA, and AAGBA.

In the upcoming sections, I tell you about the choices you have in fiber goats and some of the pros and cons of each.

Angora

Angoras have long, wavy coats, with fiber called *mohair.* (Angora fiber comes from rabbits.) They are usually white, but some breeders are experimenting with producing other colors. Their horns can be long or short and are curved or spiral, in the case of bucks. They are usually left on the goat, because they may regulate body temperature. The average adult goat produces 8 to 16 pounds of mohair each year, while kids give from 3 to 5 pounds of longer, finer hair. Figure 3-8 shows an Angora buck.

FIGURE 3-8: An Angora buck.

Angora goats came from Turkey, and their name is derived from the city of Ankara, which used to be called Angora. These goats were first brought to the United States in 1849. Since then their popularity has grown, so that now the United States is one of the biggest producers of mohair. Most of the producers are in Texas.

TECHNICAL STUFF

If you're wondering whether anyone has tried crossing Angoras with cashmeres, the answer is yes. But the resulting goats — called *Cashgoras* — were a bit of a flop, producing fiber that isn't as useful or as high-quality as the original products.

Angoras may be raised on a range, but in that case a couple of problems arise. They are sensitive to temperature and cold, and wet weather can kill them. They also are not natural mothers and so they sometimes abandon their kids. The kids often need help to get started nursing. And with twins and triplets, the bigger kids hog all the milk if not controlled, leaving the little ones nothing to eat.

Cashmere

Cashmere goats in the United States aren't a breed but a type of goat. What makes a goat a cashmere goat is that it produces cashmere. According to the Eastern

Cashmere Association, feral goats from Australia and Spanish goats in the United States are both cashmere producers. They just need to have been bred to produce the right quality of cashmere, a measurement that is determined by the cashmere industry. (The fiber has to be less than 19 microns thick.) Most of the larger cashmere breeders originally imported high-quality fiber goats from Australia to start their herds.

Cashmere goats are dual fiber/meat goats. Like their non-cashmere feral relatives, they are quite hardy. But like dairy goats (and the Wicked Witch of the East) they don't like rain and will run for shelter when it comes. Refer to Figure 3-7, which shows a Spanish cashmere goat.

U.S. cashmere goats are a little smaller than standard dairy goats but larger than Nigerians, making them a medium-sized goat. The bucks weigh about 150 pounds and does about 100 pounds at maturity. Like the Angora, they aren't disbudded, because their horns help them regulate temperature so their heavy coats don't overheat them.

The cashmere is actually their downy undercoat. Their outer coat, called guard hair, is quite coarse and isn't used for fiber.

TECHNICAL STUFF

Cashmere production in the United States is still a small market, so if you get cashmere goats, you will be part of an elite group. Breeders with large herds remove the cashmere once a year, between December and March. That's because it grows in response to light — from the summer solstice until the winter solstice. If you don't have a lot of goats, you can just comb out the cashmere when it starts to shed. The average cashmere goat produces only about 4 ounces of cashmere per year.

Miniature fiber goats

I haven't yet seen or heard of a miniature meat goat (probably because they don't make enough burger), but people seem to have a general predilection for miniaturizing animals. That tendency applies to fiber goats, as well as the dairy goats.

You find two different breeds of mini fiber goats in the United States — the Pygora and the Nigora. As you can tell by their names, both are crosses with the Angora.

Pygora

The Pygora is a cross between the Pygmy and the Angora that was started by Katherine Jorgensen in Oregon. She wanted to use the fiber for hand-spinning. The Pygora is a small, easy-to-handle, and good-tempered fiber goat. These little guys are registered by the Pygora Breeders Association (PBA), which started in 1987. A Pygora can be up to 75 percent of either breed.

Pygoras come in the same colors as Pygmies. Does weigh between 65 and 75 pounds, while bucks and wethers are between 75 and 95 pounds. The PBA requires them to be a minimum of 18 inches tall (does) and 23 inches (bucks).

Pygoras can produce up to four pounds of fleece a year, a bit less than the full-sized Angora. They are smart and can sometimes be found in petting zoos and circuses.

Some people raise Pygoras for pets or brush-eaters and never use the fiber. Because of their longer hair, these goats are more prone to lice and should be sheared twice a year. (See Chapter 18 for more about harvesting goat fiber.)

Nigora

Nigoras are a cross between a Nigerian Dwarf or a mini dairy goat (Nigerian crossed with a full-size dairy goat) and an Angora. They have the advantage of producing colorful fiber, as well as milk. The American Nigora Goat Breeders (ANGBA) was started in 2007 and several groups dedicated to the breed can be found on Facebook.

If you have trouble finding Nigora goats, you can still buy a Nigerian buck and some Angoras and start your own herd!

CONSIDERING CROSSBREEDS

Any of the goat breeds can be crossbred to each other, either intentionally or by accident. (Many a crossbreed has mysteriously appeared five months after a buck escape!) These inadvertent crossbreeds usually can't be registered without genetic testing, and so they go for a bargain. ADGA accepts and records crosses between their standard size breeds into their association as "experimental" and provide classes for them at their shows. Their offspring can be graded up to "American."

Crossbreeds also may be cheaper than purebreds if they are from farms that don't really care about the goat breed, but just want a milker or some meat.

Don't discount the crossbreeds or experimentals. They may be just what you want if you don't care about a specific breed, or if you *really* care about the breed and want one of the new, experimental breeds such as a mini dairy goat or a Nigora.

» Identifying which plants are poison

» Becoming aware of potential predators

» Protecting your goats with a guardian animal

» Constructing a milk stand for your goats

Chapter 4
Getting Your Property Ready for a Goat

Bringing goats into your life requires a fair amount of preparation. You need to make sure that you have enough space and that the space you have is safe before you introduce goats.

In this chapter, I tell you how to decide how many goats to get, and what to do to keep your goats safe after you bring them home. I talk about fencing, potential predators, and the pros and cons of various guardian animals that you can use to help protect your goats against predators.

Figuring Out How Many Goats You Can Support

One of the biggest mistakes that people make is getting too many goats. You need at least two goats so that they can keep each other company, but it's better to start slow. And you need to get only wethers or does, depending on your purpose, unless you have seriously considered the implications of getting a buck and decided that you really need one. (See Chapter 7 to find out more about getting your goats.)

The number of goats you can adequately provide space for on your property depends on the answers to a few questions:

- **How much fenced pasture or range is available to the goats?** If you live in an area where you can't let your goats *range* (roam over a large area), such as a desert area or in the city, you need about 20 square feet per adult standard-sized goat for sleeping and resting, plus another 30 square feet (outdoors, ideally) for exercise. This gives them enough space to move around and not be cramped or too confined.

  If you have a larger outdoor area in which to raise your goats — where they'll have pasture, woods, or range — you need less indoor space per goat because they only rest and sleep there. The rule of thumb is 10 to15 square feet per adult standard-sized goat.
- **How much space do you have for a sleeping area?** Goats like to sleep together in small groups, and so the actual sleeping area they need can be quite a bit smaller than their living area. If you have a building with a lot of separate pens, keep the doors open so they are accessible to all of your goats. (Of course, the herd queen and her brood will take over the best spot anyway.)
- **Will they have kids?** A standard-sized goat needs at least a 4-foot-by-5-foot kidding pen. If you have more than one doe that you want to have kid, breed them at different times, and clean and sanitize the pen between kiddings.

Making Sure Fencing Is Adequate

Goats need good fencing to keep them in and to keep predators out. You can also use fencing to protect your trees and shrubs from goats. (See the section "Making trees goat-proof.") Goats are intelligent and curious, and if they spot a weakness in your fencing, they take advantage of it and escape.

Adequate fencing means different things in different situations. If you have kids and adult goats, you need to make sure the kids can't get through the fencing and the adults can't get over it. Permanent fencing around the perimeter of the goat area has to be strong enough to last for many years, while temporary fencing that allows goats to browse in different areas only has to keep them in for a limited time.

TIP

Goats love to rub on walls, fences, and even people. If you put in new fencing, make sure that you set your fence posts deep enough so that they can withstand the weight of a goat dragging his body across the fence regularly. If you use wooden posts, they need to be at least two feet deep. If you are using metal T-posts, make sure to pound them in past the V at the bottom that holds them in the ground.

If you have an area with existing fencing, walk the fence line and

- **Inspect the fence for holes in or under it.** Patch holes in the fence and fill or block holes under it.
- **Check each fence post to make sure it's solidly set.** Replace, add a new post, or solidify the weak one.
- **Measure to see whether the fence is high enough.** The necessary height depends to some degree on the type of goat and the type of predators in your area. A 4-foot fence is adequate in most cases. If it isn't, add a strand or two of electric wire or fence it higher.

  If the area you're fencing borders a busy road, don't skimp on fencing. Make sure your goats can't get over it and cause an accident or get injured.
- **Determine whether any trees need to be fenced out or around.** Keep goats away from trees you don't want eaten or that are poisonous. See the section "Protecting Your Plants" for more information.

Running through types of fencing

If you need to put in new fencing or replace fencing, you have a variety of fencing types to choose from. The most common types of fencing that people use for goats include:

- **Field fencing:** Field fencing, also called *woven wire,* attached to metal T-posts is probably the most common type of fencing for goats. It is moderately expensive and is sturdy if installed properly. A four-foot-high field fence will keep miniature goats in but isn't high enough for a determined goat that is bigger. A strand of electric wire along the top and 10 inches off the ground usually keeps all goats in. I recommend field fencing. One downside is that it is hard to use in hilly areas.
- **Cattle or hog panels:** Galvanized cattle panel with graduated spacing (smaller to larger from the bottom) makes excellent fences for goats. The panels are 50 inches high — tall enough to keep most goats in (I say "almost," because I had a Nigerian Dwarf buck with only three good legs climb one to get to does in heat). You can add a strand or two of electric wire along the top for larger goats. To keep in miniature goat kids you may need to reinforce with chicken wire or woven wire along the bottom.

This fencing is easy to install; you can attach each panel to T-posts set 4 feet apart. The disadvantages are that the panels are very expensive, hard to work with on hills, and hard to transport because they come in 16-foot lengths and are heavy. If you can overcome these obstacles, I highly recommend using cattle panels for fencing.

» **Electric wire:** Electric wire is an excellent addition to any of the other types of fencing. The wire and insulators are inexpensive; the biggest cost is the charger and ground rod. A strand along the top helps keep predators out and goats and livestock guardian dogs in. A friend of mine uses three strands of electric wire about a foot off the ground with temporary posts in areas where goat escape isn't a disaster. The advantage is that it can be moved if you want goats to browse different areas. The disadvantage is that some goats can walk right through it with seemingly no effect. (Or they get out and then are too scared to get back in.) You must train your goats to the fence prior to using it this way. You can do this by setting it up inside a fenced area and then watching them while they test it out. Normally they learn after only two shocks.

TIP

Use a 4,000-volt charger for goat fencing. If your fencing isn't near a power source, get a solar charger. Place the grounding rod in a location that is as dry as possible. Follow the manufacturer's recommendations for grounding and charger placement.

If you want the best electric fence, ideal for keeping coyotes out, you can invest in *high-tensile electric fencing,* also known as New Zealand fencing. It combines woven wire and electric in one fence, with alternating wires electric. It is more expensive than standard electric wire.

WARNING

Avoid using barbed wire or wood fences for goats. Barbed wire fencing is less expensive than woven wire or cattle panels, but goats can get injured by the barbs. Wood fences, such as the white ones used for horses, are very expensive, don't last as long, and are inadequate for goat kids, who can get out between the boards.

Planning for gates

Plan space for enough gates in the places you need them — highly trafficked areas, near structures, and between different pastures. Make sure they are strong enough to withstand goats. This means a good, goat-proof latch. If you use a hook and eye latch, make sure that it has the extra piece that locks so it can't just be pushed out of the eye. I recommend that you install gates so they open *into* the enclosure, otherwise a goat can more easily get loose by rubbing or shoving on them.

Make the gates as high as the fence, if possible, with no gaps underneath the gate. If you will need to drive or get a riding lawn mower or tractor into the area, make sure the gate is wide enough.

You can purchase premade gates of differing widths. If you use cattle panel for fencing, you can make gates from smaller pieces. If you make cattle panel gates, overlap them with the fence to give added strength, especially in an area that houses bucks.

Protecting Your Plants

Goats don't appreciate the aesthetics of a well-tended flower garden or a fruit orchard. (They do appreciate the taste of it, though.) They are browsers, which means that they prefer bushes, trees, and woody plants. They also prefer variety in their diet and so try most of the plants that are available.

If you have a favorite lilac or rose bush, apple tree, or even a large fir tree, be ready to have it relieved of leaves, bark, and ultimately its life. If you want to keep any flowers or bushes and trees, make sure they aren't growing in an area where your goats might go. You can fence them off, or in the case of trees, you can goat-proof them. The upcoming sections give you ideas for protecting your plants and trees.

Considering which trees to protect

You need to remove any trees that are poisonous to goats and fence off or goat-proof any others you don't want destroyed. (See the section "Removing poisonous plants" to find out which common trees can be poisonous.)

If your property is large enough, you can fence off an area for your goats that doesn't have trees, or you can decide to sacrifice some trees to the goats. After all, the bark and leaves are good for them. You can even locate the pen where the leaves can still fall or branches hang over the fence so your goats can enjoy the trees without destroying them. If the area you need to fence includes trees that you want to protect, try goat-proofing the trees. The next section tells you how.

Making trees goat-proof

Goats will damage and eventually kill trees by browsing on the leaves and shoots, stripping the bark, and rubbing their horns (if they have them) on the trees. The only trees I have that seem to be immune to goat damage are old-growth Douglas firs whose bark is probably not tender enough for them. Your goats cause worse damage when they don't have access to any other plants to eat, but they enjoy tender bark and leaves even when grass and shrubs are available.

If you want to save trees that are within the goats' fenced area or just outside the fence, you need to goat-proof them. A better choice, if you can manage it, is to avoid including trees in the goats' area; goat-proofing individual trees is time-consuming and expensive.

TIP

For smaller trees or saplings, you can buy 5-foot-tall tree bark protectors from garden stores or wholesalers. These mesh or corrugated plastic tubes were designed to fit around the tree trunks to protect them from deer.

You can goat-proof a larger tree simply by wrapping it up to the level that your largest goat can reach when standing with its front legs on the tree. You can determine this height by holding a treat up next to the tree and measuring how high the goat can reach. (If your goats aren't full-grown, you have to estimate.)

WARNING

One downside to goat-proofing is that it can inhibit growth of the tree, so you need to check the tree to see whether the wrapping is too tight and rewrap it every few years.

Materials you can use to wrap a tree include

- **Plastic strips designed to cover rain gutters to keep leaves out:** I found these at a garage sale. I wrapped them around the tree and hooked them together with wire.
- **Hardware cloth, also called rabbit wire:** It is more rigid than gutter covers and can be attached to the tree or to posts in the ground surrounding the tree. However, it's expensive and doesn't fit as neatly as gutter covers. If you attach hardware cloth directly to the tree, it also will inhibit outward growth.
- **Window screen netting:** You can obtain a 1,000-foot roll for less than $100 and use it to wrap quite a few trees, depending on their diameter.

If you have only a few trees at risk of being debarked and want something more permanent, you can goat-proof trees with triangular wooden enclosures like those in Figure 4-1. A wooden enclosure requires three posts as tall as the tallest goat can reach, and wooden slats attached with screws across them. You can attach them close enough together to prevent the goats from getting their heads through the slats, or you can make the enclosure far enough outside the tree that they can't reach even if they get their heads through.

FIGURE 4-1: A tree goat-proofed with a wooden enclosure.

Protecting Your Herd

After you finally get the goats you dreamed of, you don't want to put them at risk or lose them from an avoidable cause. In this section, I tell you about some of the steps you can take to protect your herd from potential hazards.

Putting hazards out of reach

When you're checking out your property, yard, garage, or barn in preparation for your goats, reconsider items that you have stored or are in use there with an eye to goat safety. If you're going to drink your goats' milk or eat them, you are also at risk of ingesting any poison that your goats get into. Remove any items that might put a goat at risk, especially

- **Lead paint:** Goats love to chew and will invariably chew on walls, especially if you don't want them to. Lead paint is common in old barns and other structures. To be safe, assume that the paint on old walls and doors is lead-based, and don't use those areas for goats. Bare, untreated wood is actually best.

- **Railroad ties:** If you are putting up a new structure and have access to free railroad ties, don't use them. They contain creosote, which is poisonous to goats.
- **Plastic:** Keep all plastic, particularly plastic bags and plastic twine, out of reach of goats. Goats that swallow plastic can suffer from a blocked rumen and lose weight or die. (See "Hardware disease" in Chapter 11.) Swallowing plastic also causes symptoms such as loss of appetite, decreased milk production, and bloating. Be careful to properly dispose of plastic from mineral blocks or other types of feed.
- **Solvents and other chemicals:** Make sure that you have removed any old kerosene, solvents, or other chemicals that people often keep in garages or barns. These hazards can sicken or kill goats. Even those stored on high shelves within a goat area aren't safe.
- **Pressed wood/fiberboard:** This is another material that some goats will chew on, if given the chance. They'll eat holes in a building made of this material.

WARNING

Store all feed away from your goats in an area they can't access. If they inadvertently get to grain, they will eat until it's gone — and then you will have very sick or dead goats. It happened to me once, and I didn't know which goats had overeaten (except the LaMancha instigator) and so I had to treat many goats. The feeling of panic is terrible, and so is the guilt when a goat dies because of your mistake. See Chapter 11 to find out about enterotoxemia (overeating disease).

Avoiding tethering

WARNING

A tethered goat is like a piece of cheese in a mousetrap. The goat is bait for coyotes, cougars, dogs, or any predator that lives in the area. When you tether your goats, you put them at risk. Because goats are browsers, they won't stand still while on a rope. They move constantly, which means that they may circle around a tree or get hung up in a branch. If they are left on a hot day, they may get dehydrated because they are unable to get to water after wrapping themselves around an object. Goats may hang or choke themselves with a chain or rope. The snap holding them may come off or break and release them into a high traffic or otherwise dangerous area.

My neighbor had a goat that he tethered in the orchard to eat down the grass. One day he had a feeling that something was wrong and came down the hill to find the goat hung from a tree branch. I don't know how he managed to do that (but then, he *was* a goat), but fortunately the neighbor got there before the goat had choked to death.

Don't tether your goats. Instead, build them a proper fence, or if you need to move them around, use cattle panel sections or electric wire to create a barrier that you can move from place to place during the day. And supervise them or get them a guardian for protection.

Providing a safe place to bunk down

The best way to ensure that your goats are safe, especially if you don't have a guardian animal, is to make sure that they are secured in a building with no open windows from dusk until dawn. Unless you have a very heavy door, make sure the door closes and latches to prevent animals from getting in and goats from getting out.

Considering local predators

Find out about the common predators in your area when deciding how to keep your goats safe. Don't think that because you are raising goats in the city that you don't have to worry about predators. While some of the animals that we traditionally think of as predators are rarely found in the city, dogs are rampant and birds abound. Kids are particularly at risk, because they are small and lack life experience.

Here are some of the more common goat predators. If you have a goat attack, you can determine which of these predators is responsible by examining the *scat* (manure) left behind and by the focus of the attack.

Domestic or feral dogs

Dogs are the worst predators of goats, attacking and killing more often than any wild animal and doing it for fun rather than because they're hungry. Dogs go after goats individually or in packs, with pack attacks being the worst for obvious reasons.

Dogs dig under fences to get to goats or other small livestock. You can identify a dog attack because dogs usually go for a goat's hind legs and rear end. They can kill or cause injuries severe enough to create major physical problems for a goat. Goats that are attacked by dogs often have to be euthanized or die later from the trauma of being terrorized and seeing their herd mates killed.

Coyotes

Eastern coyotes hunt individually, looking for weak members of a herd; western coyotes hunt in packs. You can tell the difference between a coyote attack and a dog attack because dogs chase and try to get as many goats as they can, while coyotes, which are looking for food, will go for the throat and then try to get at a goat's internal organs. They may even try to carry the animal away for safe eating.

If you hear coyotes howling in your area, your goats are at risk if they don't have additional protection such as a guardian animal or high tensile electric fencing. Neither of these is failproof, though; I have heard of packs of coyotes luring a dog away while other cohorts go after the goats.

Cougars

Cougars hunt individually. They leave tooth punctures and claw marks on the upper torso when they attack a goat. They also have been known to drag their prey a distance away, bury it, and come back later to eat. A good livestock guardian dog will normally deter a cougar, unless it is very hungry. Nevertheless, I know of a rancher whose three Great Pyrenees guardian dogs were killed by a cougar one by one over several nights.

TIP

If you live in an area with cougars, get more than one livestock guardian dog to protect your goats.

Birds

Ravens and black vultures attack goats from time to time — generally when they are down from sickness or trying to have their kids outside. Ravens peck an animal's head and gouge out its eyes. Ravens attack in groups, which causes a problem for does trying to protect more than one kid.

The USDA recommends hanging an *effigy* (fake vulture carcass) to deter vultures. Owls, eagles, and large hawks also may bother small kids, especially if they get separated from their mothers and cry. You can prevent losses to all types of birds by making sure your goats have safe, indoor kidding pens.

Other predators

Wolves, bears, foxes, wild pigs, and even feral cats will go after goats if their regular food supply is disrupted. You only have to read the newspaper to realize that humans are also predators on goats — some rustling for food, but others killing for the fun of it, as a prank, to get even with the owner, or for some other misguided reason.

Using guardian animals for security

Many goat owners keep livestock guardian dogs, donkeys, llamas, or alpacas with goats as full-time guard animals. Guardian animals can add a substantial cost in terms of training and upkeep, but they may be well worth the effort and time if they work out.

TIP

Try to get a guardian animal from a breeder who has used the animals for this purpose and can vouch for (but not guarantee) their pedigree, training, and temperament.

Livestock guardian dogs

Livestock guardian dogs (LGDs) were bred and have been used for thousands of years to protect goats or sheep in Europe and Asia. They live and bond with the goats, are aggressive toward predators, and are focused on the job. These dogs are traditionally white, which enables them to blend in with the sheep flock and be distinguished from predators. Of the many breeds of livestock guardian dogs, the Great Pyrenees is probably the best known. Other common livestock guardian breeds include

- Akbash
- Anatolian
- Komodor
- Kuvasz
- Maremma
- Ovcharka
- Shar Planinetz
- Slovak Cuvac
- Tatra Sheepdog

LGDs live with the livestock they protect yet have a relationship with their human caretakers. They are large, aloof, and can be intimidating to strangers. If you plan to get one, expect to do a lot of training and supervision because they don't reach maturity for about three years. You have to be the alpha member of the group so the dog knows you are in charge. Neutering or spaying a guardian dog also helps ensure that it will be a successful goat-guard.

An LGD probably won't work out well for you if you're raising goats in an urban area. They warn predators to stay out of the area by barking, mostly at night. Make sure you're aware of any noise ordinances and know your neighbors and whether they will take issue with a barking dog before getting an LGD.

Some livestock guardian breeds also have a tendency to roam, which may cause problems if you have neighbors who don't want your dogs on their property or you have a road with traffic nearby. Because I have only a small property, I have had a problem with several of my LGDs roaming. Luckily, with the first one, my neighbors appreciated having a guardian dog and welcomed her patrolling their property and the woods behind our land. The second one was another story: He had a bad habit of going in the road. At age 12, he developed dementia and had to be put down when he started wandering in the road constantly. My third LGD is a female Maremma who is completely bonded to her herd. She wouldn't consider getting out and roaming. I made sure to purchase her from breeders who had both parents on their property, and they were proven to be good guardians.

If you have a large herd or a large area where your goats browse, one guardian may not be enough to protect them. Guardian dogs often act in teams to protect against packs of predators. Make sure before you get your dog that it's from a stock bred to be guardians for sheep or goats.

LGDs live to be only 7 to 9 years old because of their large size. Plan to add new dogs periodically as your protectors age.

Don't buy a herding breed such as Australian shepherd or border collie to guard your goats; they aren't qualified. Their job is to herd, and you may have a problem with them chasing goats. That isn't to say that some haven't been successful, just that they are unlikely to do a good job guarding and protecting goats.

Donkeys

Donkeys have been used for hundreds of years to guard sheep and other herd animals. They're very intelligent and have good hearing and eyesight. They work better alone and don't like dogs, so they can't work as a team with an LGD. Donkeys' dislike of dogs also makes them effective against coyotes (which aren't dogs, but look like them). Miniature donkeys are also not good candidates, because they're small enough to need protection of predators themselves.

Because donkeys are naturally herd animals, if they're bonded to the goats they can be counted on to stay with them most of the time. Ideally, you get a guardian donkey at birth or as soon as it's weaned to make sure it bonds with the goats. Because they eat the same food as the goats, donkeys also will want to stay with the herd after they realize that's where the food is.

When a guardian donkey becomes aware of a predator, she situates herself between the intruder and the herd and brays loudly. If the animal doesn't leave, she chases it, and if that doesn't work, she attacks by rearing up on her hind legs and coming down on the predator with her front feet.

TIP

Intact *jacks* (male donkeys) are too aggressive to be effective guardian animals, and so you need to get a *gelded* (neutered) jack or a *jenny* (female) if you want to use a donkey as a guardian. You'll also be better off finding one that has been bred as a guardian. Donkeys not raised with goats or other livestock may be aggressive toward your goats, especially at feeding time. If you get one that wasn't previously bonded to goats, be sure to introduce her slowly and supervise them closely for several weeks. When you purchase a guardian donkey ask for an agreement to exchange the animal if it doesn't work out.

Llamas and alpacas

Llamas and alpacas are good guardian animals because they bond quickly to goats and also eat the same feed. Castrated males make the best goat guardians. Males can injure goats by trying to mount them and can be too aggressive toward humans as well. I had a female llama that I tried to use as a guardian, but she never really understood the job. She loved to chase goats for sport. She had been handled by humans for too long and didn't get an early start on the job.

TIP

Because they're smaller, alpacas are not the ideal guardians, especially if large predators are in the area. A neighbor of mine had a series of alpacas killed by a local cougar.

I also disliked the llama trait that people are so pleased with: always using the same spot to relieve themselves. Unfortunately, the spot my llama chose was right in the middle of the barn. Rather than go out in the pasture, if she felt the urge she came into the barn and let loose.

Unlike dogs, llamas work better as guardians when they're alone instead of in a pack. A llama and guard dog combination can be trained to work cooperatively, though.

Llamas need strong fences to help them do the job. If a guardian llama can't scare off a dog or coyote with his aggressive attitude, the predator may kill him.

Removing poisonous plants

Goats ignore poisonous plants most of the time, but because of their need to browse, they may try them just for variety. Whether goats that eat a poisonous plant show signs of poisoning depends on how much of the plant they eat, what part of it they eat, the condition of the plant (fresh or dried), the time of year, and the size and health of the goat. They may have no signs of poisoning, or they may get very sick and even die. Chapter 11 tells you how to identify signs of poisoning and what to do for a poisoned goat.

Some of the common poisonous plants that might grow in your pasture or backyard include:

- Weeds
  - Bracken fern
  - Buttercup
  - Common milkweed
  - Foxglove
  - Lantana
  - Locoweed
  - Poke weed
  - Spurge
  - St. John's Wort
  - Water hemlock and poison hemlock
- Trees
  - Cyanide-producing trees such as cherry, chokecherry, elderberry, and plum (especially the wilted leaves from these trees)
  - Ponderosa pine
  - Yew
- Cultivated plants
  - Azalea
  - Kale
  - Lily of the valley
  - Oleander
  - Poppy
  - Potato
  - Rhododendron
  - Rhubarb

Many landscaping plants are poisonous, and a few are so deadly that even a few leaves can make your goat extremely sick. Don't believe the old wives' tale that goats always know what is poison or not. Before you bring your goats home, check your yard for these plants. The best resource for doing so is *A Guide to Plant Poisoning of Animals in North America* (Teton New Media), by Anthony P. Knight and Richard Walter. You can find many chapters of it online at the International Veterinary Information Service (`www.ivis.org`). I can provide only limited information here, and so I recommend you check out the website and the book to make sure you're familiar with the poisonous plants in your area.

If your goats can get their heads through a fence to the neighbor's yard, make sure that poisonous plants aren't growing within reach there.

If you find any of these plants, either remove them or make sure that your fencing will keep your goats away. If the poison plant is a tree, make sure that the leaves won't fall into the pen in the autumn by either removing the tree or situating the pen far from the tree. Dried leaves can be the most deadly part of the tree.

You usually don't need to freak out if one of your goats eats a little taste of any of these plants or trees, but you do need to keep an eye on him in case he shows signs of sickness. (See Chapter 11 for more about poisons.) I had a rhododendron plant in my front yard for years, and when I let the goats out they sometimes took a little taste. I found it nerve-wracking and finally removed the plant, but none of my goats ever showed signs of poisoning.

I recommend that you talk to your neighbors about poisonous plants and ask them not to throw their garden trimmings into the yard as a treat for your goats without asking first. I lost a goat kid to oxalate poisoning from kale that a friendly neighbor had given my goats.

Building a Milk Stand

One essential item for goat owners is a milk stand, sometimes referred to as a *stanchion*. This device enables you to secure your goat not only for milking, but also for doing all the routine care and important tasks that I tell you about in Chapter 10.

You can purchase a metal milk stand at a livestock supply store or make a wood milk stand yourself, if you have the skills and equipment. Caprine Supply and other online livestock product suppliers sell the headpiece (technically, the headpiece itself is the stanchion) separately. For bigger goats, there is one that the head rests on instead of having a locking mechanism.

The simple, wooden milk stand in Figure 4-2, is for miniature goats and is easy to make. If you want to make one for a larger breed, you can lengthen the plywood and the side boards and make the legs longer. A similar milk stand can also be found online for only about $100.

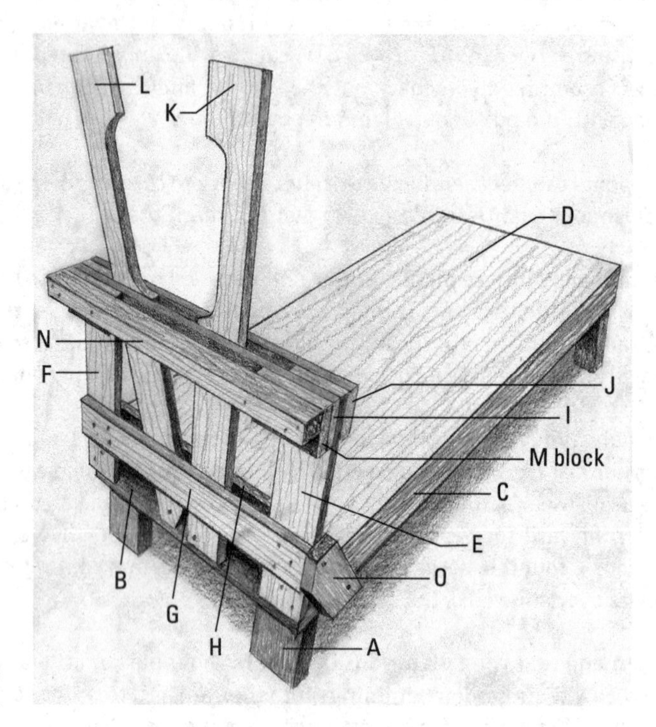

FIGURE 4-2: A milk stand.

Before you dive into hammering, gather your materials and equipment. You'll need the following from your local lumberyard:

- One 2 x 4 measuring 8 feet in length
- One 1 x 6 measuring 10 feet in length
- Two 1 x 4s measuring 10 feet in length
- One 2 x 2 measuring 8 feet in length
- One sheet of ⅝-inch plywood measuring 36 x 21½ inches

Pick up the following hardware:

- One box of 100 1-inch deck screws
- One box of 50 2-inch deck screws

- One ¼-20 2½-inch bolt
- One ¼-20 lock nut
- One 4-inch hook-and-eye latch

Finally, make sure you have the following tools handy:

- Drill
- Drill bits: ⅛ inch and 5⁄16 inch
- Saw
- Sandpaper (and a sander, if you have one)
- Pencil
- Square
- Tape measure
- Screwdriver
- Clamp

The following sections cover all the steps for building.

Cutting the lumber into parts and marking the pieces

Start by doing the following:

1. **Cut the 2 x 4 into four 14-inch-long leg pieces, and label each of these pieces A.**
2. **Using the remainder of the 2 x 4, cut a 20½-inch piece for the feed holder, and label the piece N.**
3. **Cut the 1 x 6 into two 20-inch end pieces and label each of these pieces B.**
4. **Using the remainder of the 1 x 6, cut two 36-inch side pieces, and label these pieces C.**
5. **Cut one of the 1 x 4s into four 21½-inch pieces, and label these pieces G, H, I, and J, respectively.**
6. **Using the remainder of the 1 x 4 from Step 5, cut one 18½-inch piece and label it F.**

7. **Using the remainder of the 1 x 4 from Step 5, cut two 4½-inch pieces, and label these pieces M.**
8. **Cut the second 1 x 4 into two 36-inch pieces, and label them K and L.**
9. **Using the remainder of the 1 x 4 from Step 8, cut one 18½-inch piece and label it E.**
10. **Using the remainder of the 1 x 4 from Step 8, cut two 12-inch pieces for braces and label it O.**

You're done cutting the lumber!

Attaching the legs to the base

Lay one end piece (B) on your work surface and screw two of the legs (A) to the inside of one end of the end piece (B) using 1-inch screws. Make sure that the screws are flush with the top and side of the end piece. Repeat with the other two legs and end piece.

If you have a partner to hold the pieces, it will go faster. If you're working alone, use a clamp to secure the pieces before screwing them together.

Attaching the side pieces to the base

Screw the side pieces (C) to the end pieces (B) using 2-inch screws. The legs and ends should be on the inside of the side pieces.

Finishing the platform

Turn the structure so it's standing upright on the legs. Lay the plywood across the top lengthwise, making sure that one end of the plywood is flush with the end of the base. When you're sure it's square, drill holes with the ⅛-inch drill bit. Screw it all together with 2-inch deck screws.

It's easier if you drill ⅛-inch starter holes first.

Preparing the stanchion

To prepare the stanchion, follow these steps:

1. **With a tape measure, measure over 2½ inches from the top corner of piece F, and mark with a pencil.**

2. **From the same corner, measure down the side 5¾ inches, and make a mark across the piece.**
3. **Using your square as a ruler, draw a line from the first mark to the second mark.**
4. **Cut with your saw along the line.**
5. **Place the stanchion supports (E and F) on a table.**
6. **Place piece H flush with the top and side and place piece G across the lines you drew, 4 inches across.**
7. **With the cut facing inward on the piece labeled I, predrill holes for 3 screws at each end of pieces G and H.**

Assembling the stanchion

To assemble the stanchion, follow these steps:

1. **Turn the stanchion upside down, with piece H on the top, and place pieces I and J on top, aligned with G and H in the same location as G1 and G2.**
2. **Drill holes with your ⅛-inch drill bit and screw it together on both sides.**

Creating neck pieces

To create the neck pieces, follow these steps:

1. **Take pieces K and L and mark a line across each piece at 4 inches and 16 inches from the top.**
2. **Mark a vertical line between the two marks you made in Step 1, 1½ inches from one side.**
3. **Using your saw, make a cut that curves from the top line, horizontally, then down the vertical line, and then curving into the bottom line horizontally.**

   TIP

   A jigsaw works best for this.
4. **Measure over from one end of piece J and piece I to the center and mark the center line with a pencil.**
5. **Measure ½ inch over from each side of the center line.**
6. **Measure and mark 3½ inches from the inside bottom of piece L.**

7. **Measure up from the bottom and mark at 2¾ inches.**
8. **Draw a line between the two marks and cut this angle.**
9. **Sand these two pieces so they're smooth.**

Adding the neck pieces to the stanchion

To add the neck pieces to the stanchion, follow these steps:

1. **Place neckpiece L between pieces I and J on the stanchion support ½ inch from the center and extending 4 inches below the bottom piece I, with the cutout area facing inward and flush with the bottoms of pieces E and F.**

   Clamp, if needed, before proceeding.

2. **Place piece K 1 inch from piece L and 4 inches past the bottom of piece I, parallel with piece L.**

   The cutout part should be facing toward the inside. Make sure piece L can move smoothly between pieces I and J. If it can't, remove it and sand some more; then repeat this step.

3. **Measure over from one end of pieces J and I to the center and mark the centerline with a pencil.**
4. **Measure ½ inch over from the center and mark again.**
5. **With piece L lined up correctly and the pieces clamped (or with a helper holding them in place), drill through pieces I, L, and G with the 5⁄16-inch drill bit.**
6. **Drop a bolt into the drilled hole, and add the hex nut to the other side to hold it in place.**
7. **Add a support block (M) to each end of piece H, making sure that the side and top align.**
8. **Use your ⅛-inch drill bit to drill starter holes, and then screw them together with 1-inch deck screws.**

Adding the feeder attachment

To add the feeder attachment, place piece N across the center of the upper M piece, with the ends aligned and flush with the top of piece M. Predrill holes and use 2½-inch deck screws to attach. You can add a plastic feeder in the front of this.

Attaching the stanchion to the base

Sand any rough surfaces of your structure until they're smooth and the corners are rounded.

Attach the stanchion to the base, with the feeder attachment on the outside and with the bottom aligned with the bottom of piece I, sitting on the edge of the base. Predrill holes and attach with 2-inch deck screws through pieces E and F and K.

Don't put a screw through piece L. It has to move freely so your goats can get their heads in and out.

Prepare the O brace by following these steps:

1. **Measure 2 inches and mark from the left corner of one end.**
2. **Draw a line from the right corner of the same end diagonally to the 2-inch mark.**
3. **Measure 11 inches up from that right corner and mark.**
4. **Measure across the other end from the same side 1¾ inches and mark.**
5. **Draw a diagonal line from the 11-inch mark to the 1¾-inch mark.**
6. **Then measure 10 inches from the original left corner and mark.**
7. **Draw a diagonal line to the 1¾-inch mark.**
8. **Saw on these lines.**
9. **Use the cut O piece as a pattern to draw and cut the second O piece.**

Attach braces O with 1-inch screws on each side to the bottom end of the stanchion and flush with the bottom of the side piece of the base.

Making final adjustments

Add a hook-and-eye latch to the inside top of the neck assembly (above the cutout). For bucks or goats with thicker necks that may not fit, you can use a bungee cord wrapped around the neck assembly to hold it together.

To make it easier to keep the milk stand clean, you can attach an inexpensive piece of flooring vinyl to the top of the platform with metal edging if you like.

2

Bringing Your Goats Home

IN THIS PART . . .

Give your goats a comfortable place to sleep.

Decide what and how to feed your goats.

Choose your goats, buy them, and bring them home safely.

Understand and shape your goats' behavior and stimulate your goats' minds.

Care for your goats with everything from basic grooming to disbudding kids to castrating bucklings and more.

» Building a sleeping shelf

» Keeping their home clean

» Controlling flies and other pests

Chapter **5**

Home Sweet Homestead: Sheltering Your Goats

Goats need a safe, clean place to hang out, sleep in, or retreat to when the weather is too hot, cold, windy, or rainy. The kind of shelter you provide varies according to your climate, how many goats you need to house, and what kinds of predators you need to protect them from.

In this chapter, I give you some ideas for shelters for your goats, tell you how to build a sleeping shelf in an already existing structure, and share some ways to keep the areas clean and disease-free.

Outlining Shelter Types

How elaborate your goat shelter is depends on where you live, what structures are available, how much you can afford, and how many goats you plan to have in the shelter. In mild weather, meat goats need only a basic shelter to protect them from sun and rain. Some of the things to consider when deciding where your goats will live include:

- **Flooring:** Dirt or gravel floors are best, although some people prefer wood. Dirt absorbs urine, and both gravel and dirt, when covered with straw, are warm. Avoid concrete because it's cold and hard on the goats' bodies, although it's easier to clean. A method of flooring that is becoming more popular is the slatted deck, which I tell you about in "Creating an Outdoor Shelter," later in this chapter.
- **Bedding:** Unless you have a slatted floor, you need to use some sort of bedding for warmth and comfort. (The upcoming section "Using and maintaining bedding" gives you more detail about bedding.) If you do have to use a concrete floor, make sure to put down 3 to 4 inches of wood shavings or straw to insulate the goats. You can also use stall mats, which make cleanup easier.
- **Dimensions:** Consider the height and width of the shelter. Goats need to have from 10 to 15 square feet of housing if they also have an outdoor area. When building, think about how easy it will be to muck out old bedding. Having to bend over or stretch a long way while mucking is uncomfortable and hard on your back, so if the goat shelter you build won't be taller than you are, don't make it too deep.

TIP

- **Climate:** An open shelter may be fine in a mild, dry climate with good fencing from predators, but it won't work in an area of heavy snow and wind or the open range. Make sure that your building is in an area with good drainage and, if it is open, that it faces away from the prevailing wind.
- **Herd composition:** If you have or plan to have a lot of goats, you need to make sure you have a large enough structure or plan to build more shelters over time. Groups such as bucks, does, and kids to be weaned need separate housing areas.
- **Storage:** Remember that you need to have an accessible place to store feed and goat-care tools. The area where you store feed must be inaccessible to your goats.
- **Access to water:** Having easy access to water for your goats is nice, but it isn't critical in most cases. Just remember that if you don't have a water source close to your goats' area, you'll be hauling water every day for a long time.
- **Other considerations:** Regardless of the breed, you need an area for doing routine care, such as hoof trimming or clipping. If you are raising dairy goats, you can use the same space for milking. In addition, if you are planning to breed your goats, you need kidding pens, which can be 4 feet by 5 feet; the

number that you need depends on how many goats are kidding at one time. I usually leave my goats with their babies in kidding pens for two to three days, and because I have only two pens, I make sure that I stagger their breeding so they each have a place to go.

Using an existing building

Consider buildings that are already on your property and may be feasible for housing goats. You can remodel a chicken house or other farm building or even use a prefab garden shed for a goat house. Many people house urban goats in a section of garage that opens into the backyard.

Check the building for ventilation — condensation buildup on the inside walls will tell you that ventilation is not adequate. Adding a window may solve the problem; goats need air coming in above them for proper ventilation.

If you have Pygmies or Nigerian Dwarves, you can use the larger dog igloos or dog houses for shelters. Kids love them and often snuggle up three to four per house. Make sure you have enough to go around, though, because as they grow they need more space.

TIP

If you have a larger building, such as a barn, that you can use for a goat shelter, think about how to organize the space before you start remodeling or arranging. Consider where you can trim hooves and give injections and, if you have dairy goats, where you'll milk. You also need to think about grain and hay storage. If you plan to breed your goats you also need a kidding pen. When you go to farms to look at goats, also look at their setup and freely borrow good ideas.

Putting up a shelter

Goatkeepers have come up with a lot of different ideas for goat shelters. These can range from a "Taj Mahal," if you have space and a lot of money to spend, to a very simple shelter when you don't have land or money.

Before you decide on a shelter, find a flat, dry area where it will sit level. Don't plan a shelter next to a fence, or your goats will soon be on the other side of the fence. They love to jump on things!

Here are some ideas for simple economical goat shelters that you can build or assemble:

- **Steel Quonset-style Port-A-Hut (www.port-a-hut.com):** The farrowing size is fairly inexpensive (about $200). The smaller ones come assembled, while the larger ones need to be assembled. They come with stakes and need to be fastened to the ground so a strong wind won't blow them over.
- **Cattle panel and tarp Quonset-style hut:** This kind of shelter can work well for meat goats in a milder climate. It is open on both ends. Because the heavy cattle panel is strong enough to withstand snow, this shelter could work in harsher climates if you build it next to a barn as an adjunct shelter. Directions for building one are at http://goatseeker.com/guides/quonset-hut-goat-shelters.

WARNING

  If your goats are climbers, be aware that they can get injured if they break through the tarp.
- **Dog run:** A dog run works well for a few small goats in a backyard. You can purchase a cover made with a tarp and in colder weather you can put tarps all the way or partly around it. Or you can put in a dog house for sleeping quarters. If it is covered on top, and also because of its height, it provides nighttime security after you latch the door because other animals can't get over and into it.
- **Wood frame shelter:** You can make a wood frame shelter of any size and use metal or regular shingle roofing. As long as the area has proper drainage, you don't need to put a floor in the shelter. Just cover the dirt with plenty of bedding. You can make this kind of shelter with a door, partially enclosed on one side or open on one side. Instructions for building a 10-foot-by-10-foot shelter are at www.ehow.com/how_4550264_build-goat-barn.html.

Providing a Safe, Cozy Place for Goats to Bunk

Goats sometimes choose to sleep outside, depending on the weather. But they need to have access to a secure indoor area to sleep in, and that's where they'll sleep most of the time. As long as they have enough bedding (and at least one more goat to snuggle with) they can sleep on the ground. But goats prefer to be off the ground so they can stay warmer and out of the way of the critters that invariably populate the areas where livestock are housed.

You meet that desire by building them a sleeping shelf. My goats' first sleeping shelf was an old church pew, but after several years, the seat disintegrated and we had to add a new wooden seat to make it functional. After I moved some of my goats to another barn, I noticed that they seemed afraid to sleep in their stall. Then I noticed the signs of a lot of rat traffic. Since we built them a sleeping shelf they've slept in the stall every night.

In this section I tell you how to build a corner sleeping shelf like my goats use.

Building a sleeping shelf in an existing barn

You can build a sleeping shelf for your goats in an existing barn, garage, or other building. You need very few materials to build it — in fact, you might already have the necessary wood left over from another project.

I give you directions here for a triangular corner shelf (see Figure 5-1) that we built in an old barn on my property. Although the directions are for this specific shelf, you can modify them to fit your particular building, or you can even make one along a full side of your building. This shelf is set 14 inches from the ground, which is a good height for miniature and full-size goats. These directions assume that the corner of your barn is square.

FIGURE 5-1: A corner sleeping shelf is easy to build.

An added benefit of a sleeping shelf like this is that kids like the safety of an enclosed area where they can't get stepped on, and they curl up underneath it when they aren't sleeping with their moms.

First, gather your tools and materials:

- One piece of ⅝-inch plywood (4 feet x 8 feet)
- Four 8-foot 2 x 4s (one for each side brace, one for center braces and one for front)
- One pressure-treated 28-inch 4 x 4
- Two dozen 1¼-inch screws or nails
- Two dozen 2½-inch screws
- Pencil
- Level
- Circular saw
- Screwdriver (preferably battery- or electric-powered)
- Hammer

Here's how you build your shelf:

1. **Cut the plywood into a triangle with one 8-foot side and two 68-inch sides.**

   Place the plywood with 8-foot side across and measure and mark the center of the top (4 feet). Draw a line from each bottom corner to the top center line. Each line will be 68 inches long.

   Cut the plywood with your saw along the two lines into three triangles. The center piece will be 8 feet on the bottom (full length of the plywood) and 68 inches on each side.

   TIP

   You can put the two remaining pieces together to make a second corner sleeping shelf, if you want. The pieces you cut from in Step 3 brace them to make them sturdy.

2. **Cut your first two 2 x 4s.**

   Cut 65 inches from one 2 x 4 and 63.5 inches from another, making a 45-degree angle at one end of each.

3. **Cut your third 2 x 4.**

   Cut two 42.5-inch pieces with one end of each at a 45-degree angle. Fasten them together lengthwise with five 2½ inch screws, making sure that the longer points of the angles are together (see Figure 5-2).

FIGURE 5-2: Sleeping shelf frame assembly.

4. **Cut your 4 x 4 in half, creating two 14-inch pieces.**
5. **Measure and draw a level line 14 inches from the floor on the walls where you will set the frame of the sleeping shelf.**
6. **Using two screws or nails, attach the 63.5 inch piece of 2 x 4 (left side) to the 65-inch piece of 2 x 4, making a V.**

   The flat end of the 65-inch piece should be level with the side of the 63.5 inch piece (refer to Figure 5-2).

7. **Attach the 8-foot 2 x 4 to each end of the V piece with two 2½-inch screws on each side through the 8-foot piece.**

   You now have a triangular frame.

8. **Attach the half-inch support piece to the frame.**

   Mark the center (4 feet) of the long piece of your triangular frame. Attach the combined 44.5 inch piece to the frame with two 2½-inch screws or nails on the pointed end and four screws or nails on the flat end, with the 45-degree angles abutting the two shorter pieces and the flat end abutting the center point of the front piece. The center of the two pieces should line up with your center mark (see Figure 5-2).

   TIP

   If you want, you can add additional pieces of 2 x 4 to the frame midway between the center support and the side to increase the support.

9. **Place one of the 14-inch 4 x 4s upright on the top corner of the frame so that it is perpendicular to and flush with one side. Attach it to the frame with approximately six 2½-inch screws at angles starting at approximately one inch from the end of the 4 x 4.**

10. **Repeat Step 9 for the front support, with your support piece upright and flush with the front of the bottom of the frame.**

    For additional strength, you can add additional floor supports on each side of the frame.

11. **Turn the frame over and place in the corner of your building. Check it with a level, then securely attach with a 2½ inch screw every 10 inches or so in solid wall.**

    If your walls are not square, use shims between the frame and wall when attaching the frame.

12. **Attach the piece of plywood to the top with 1¼-inch screws every 8 inches or so.**

You won't even have to invite your goats to get up on the shelf. Within a few minutes after you finish it, they'll be jumping on the shelf and competing for the choice spot.

Using and maintaining bedding

Most people put bedding on their barn floors to make it more comfortable for goats and to absorb waste products. Another option is a slatted floor (see "Selecting flooring," later in this chapter).

You can put bedding down directly on your barn floor, which may be dirt, cement, wood, or gravel, or you can buy rubber stall mats. These are rather expensive and can be purchased at a farm store. The good news is that they only have to be purchased one time. Stall mats make cleanup a little easier, and if you have a cement or gravel floor, they make for a softer and warmer bed for goats. If your floor is wood, the stall mats will help protect it from composting along with the muck.

You have several options for bedding:

- **Straw:** Straw is the bedding I like most. It's easy to store because it comes in bales, and it's inexpensive. I prefer wheat straw to rye or other straws because
    - It's easier to muck out when used.
    - It's less dusty.
    - The goats like to eat it when it's fresh.
- **Wood shavings:** Depending on where you live, wood shavings may be a better option. If you're in a region with little rain, you won't have a problem with storage, because you can even keep them outside.

When I had guinea pigs I got free wood shavings from a boat-building apprentice program. Now I have to buy them by the bag, which is expensive. If you buy them loose by the truckload, storage may be a problem, but they are less expensive. I use wood shavings under the straw in my kidding pens for extra absorption.

- **Wood pellets:** Wood pellets absorb urine and odors but are too hard and uncomfortable by themselves for goats to use as bedding. They also are expensive. I sometimes use them under my straw bedding in highly trafficked areas for extra absorption. They break down and turn into sawdust over time.

When the bedding gets saturated with spilled water, urine, and feces it becomes a perfect breeding ground for flies and parasites, and must be *mucked out.* Mucking out a barn involves removing all the used bedding down to the floor and replacing it with clean bedding. This helps prevent the spread of parasites and other problems.

How frequently you need to muck your barn depends on the size of the area and how many goats you have. In the winter, if you live in a cold area, you can allow the muck to build up and add new bedding to the top. This provides extra heat for the goats from the composting bedding under the fresh layer. In the summer, you may be able to get away with mucking only once a month or so, if your goats spend more time outdoors.

If you have a large area to be mucked and are lucky enough to have a tractor or similar equipment, you can use that. But if you're like most of us with only a backyard or a small homestead, you'll have to muck by hand. To muck a barn by hand you need

- Gloves
- Muck boots or old shoes
- A pitchfork
- A wheelbarrow
- A strong back and arms
- Country music and beer (optional)

TIP

Pace yourself. If you have a large area, start on one side and finish that first. You can do the other half the next day. I like to have one or two people removing the used bedding and one running the wheelbarrow.

Use gloves to prevent blisters and muck boots to keep your shoes and clothes clean. If the used bedding is very deep, to save your back, take it off in layers with your pitchfork rather than trying to lift huge chunks.

Move all the used bedding to a single pile in a place where goats won't be tempted to play on it. They can get parasites from it. The pile may seem high at first, but with rain and time, it will shrink down to nice compost. Some people cover their muck pile with a tarp to aid in composting. Because goat manure doesn't burn plants like chicken manure does, you can put it directly on the garden, if you choose.

TIP

If you live in an urban area, be sure to check your city regulations before making a large compost pile, especially if it can be seen by neighbors or from the road.

After you remove the used bedding down to the floor, spray with vinegar to neutralize the urine and sprinkle agricultural lime, wood shavings, or wood pellets on it, if desired, and add fresh bedding.

Put 3 to 4 inches of bedding down on a concrete floor; less on wood or dirt floors — I put down an inch or two of straw and add more as time goes by. When you have an idea how much hay your goats are wasting (remember, after hay is on the ground, goats see it as bedding and not food), you can strategically add bedding in areas that are used more. Eventually the wasted hay will fill in the gaps.

Creating an Outdoor Shelter

If you want to shut your goats out of the barn on nice days (believe me, your barn will be much cleaner if you do) or if you have a large acreage that goats can roam, you can build an outdoor shelter somewhere in your pasture. You can design it so you can hang a hay feeder and water buckets for when they want something different than pasture or browse.

Protecting your goats from the elements

A structure can be accomplished in a variety of ways, from using a truck canopy as a roof on a base structure to a two- or three-sided building with low walls. It doesn't have to be anything fancy. The goal is to protect your goats from sun and rain when they aren't out browsing. They'll also be healthier from the fresh air they get than they would inside a barn.

Figure 5-3 shows a simple three-sided shelter made from wood pallets covered with plywood. You can get free pallets from factories, building sites, large farms, and farm stores. You need to purchase two-by-fours, plywood, and roofing materials. The shelter in Figure 5-3 has a wood floor and a roof made from leftover metal roofing. Two to four medium-size goats can sleep comfortably in this shelter out of the rain or sun. Slatted flooring (see the next section) is a good option for this kind of shelter.

FIGURE 5-3: Old roofing material and pallets can become an outdoor shelter for goats.

Make sure that when you consider where to locate it and which direction you want the opening(s) to face that you think about which way the wind tends to blow. It's okay to have the wind circulating across the structure, but you don't want to construct it in a way that will put them at risk of getting soaked and cold if an unexpected storm blows in. I know my goats *hate* rain.

Selecting flooring

You may want to start out with pasture grass as the flooring, but if your goats are going to spend a substantial amount of time under the structure during their siestas or in inclement weather, you'll want to create an easy-to-clean floor. This is where *slatted flooring* comes in very handy.

The idea is to create a slatted deck with treated wood that can be easily maintained, is inexpensive, and meets the goats' needs. The slats should be spaced ½ inch to ⅝ inch apart to be most effective and safe for goats. This style of flooring allows manure and urine to go through the slats and end up on the ground. The manure, along with hay that falls through, will also start composting, creating a little heat during cooler days.

Cleanup is easy because the manure doesn't get smashed down by goats walking on it, so it can be easily scooped up. To clean up, you simply tip the flooring up and scoop out the manure and any hay that may have fallen through, and then spray the bottoms with vinegar.

Slatted sections that are 4 x 4 feet in size are easy to tip up and clean as needed, which can be as frequently as every couple of weeks to every six weeks or so, depending on the amount of use they're getting. They work best when you clean the tops every day by raking with a metal rake to keep them clean and avoid buildup of manure that hasn't gone through the slats.

Keeping Your Goats and Their Living Space Clean

Keeping goats healthy includes keeping their living space clean. Luckily, their pellets are relatively dry and don't cause much odor or mess when they are healthy. You can't prevent a goat's area from getting dirty, because they don't normally think about where they relieve themselves. Still, you can do some things to keep their space cleaner.

If you don't lock your goats out of the barn, you can still do some simple day-to-day cleanup, such as sweeping goat pellets off the sleeping shelves, scrubbing buckets, and spreading the wasted hay that accumulates by hay feeders to cover fresh urine and feces.

Controlling flies and other bugs

Flies, mosquitoes, and other bugs are part of raising livestock, like it or not. But you do have some options for controlling these critters.

The cheapest way to minimize pests is to keep the barn or goat shed dry and remove wet bedding weekly during warm or hot weather. Pay special attention to the areas around water buckets, where the flies are more likely to breed.

You can also build a bat house for the outside of the barn or on a nearby tree. A bat house is a simple structure that simulates the narrow spot in a tree where bats like to live. You can buy one or find plans for building one from Bat Conservation International or on Amazon.com. Bats eat mosquitoes and other insects, and so having them around is beneficial to a farm. The only disadvantage to having bats is that they carry rabies, which a curious goat can get from inspecting a crazy-acting bat.

To discourage mosquitoes, walk around your property to check for and eliminate standing water, where they can breed. If you have water that continues to build up, just pour some vegetable oil on the top to stop the mosquitoes from breeding. Or you can buy a product called Mosquito Free Water from livestock supply companies; it's safe and can be sprayed on standing water. It breaks the surface tension of the water, making it impossible for mosquitoes to land.

Some people use insecticides to control flies and other pests, and you can find many of them on the market for just this purpose. I avoid them because the chemicals are poison that can harm you and your goats. I don't really want insecticide in my milk or meat.

Flies have a two-week life cycle; the larvae develop in the first week, pupate in the second week, and then emerge as adult flies. Your fly problem can quickly get out of hand in this short time. After more than a decade of raising goats, I have tried virtually every non-toxic method of fly control. These include:

- **Fly parasites:** You can buy these tiny wasps, which feed on fly larvae, from biological pest control companies and some livestock supply catalogs. You get them in regular shipments through the mail and put them out through the summer near wet areas, where flies are more likely to hatch. The parasites are about the size of a gnat and virtually unnoticeable. Some people swear by this method, but I have found it only somewhat helpful. It is also expensive.
- **Fly strips:** These are inexpensive and amazingly effective. You unwind these sticky strips and hang them with tacks around the barn. When the flies land on them they get stuck. Mr. Sticky is the most effective brand.
- **Mr. Sticky mini roll:** This product is superior to fly strips because the roll lies horizontally rather than vertically, and so flies are more likely to land on it. The mini kit is perfect for a small barn. It is an 81-foot reel that you attach to the walls across a stall and roll up as it fills with flies. You can also buy a 1,000-foot roll. I think these are the best non-chemical fly control ever. They cost more than fly strips but do ten times the job. I love reeling in all those dead flies.
- **Fly trap:** You can find various fly traps on the market. One is a plastic bell-shaped trap that you fill with stinky bait and hang from the ceiling. Flies are attracted to the smell, crawl in, and then can't get out. These smell bad, so you need to place them where you won't accidentally walk into one.

» **Citronella fly spray:** I also use citronella spray, which is made for horses. You can spray it directly on goats when they are being bothered by flies. I use this only when the deer flies or horse flies are particularly aggressive.

You can try any of these methods, or a combination, to see what works best for you.

Feed storage and ratproofing

Always make sure to store your hay, grain, and other feed in a location that your goats can't reach. You also need to keep grain, opened bags of Chaffhaye, beet pulp, and other supplements out of the reach of rats, mice, and other vermin. (Chapter 6 tells you about feeding your goats.)

TIP

You can store hay in a loft, in a closed stall with a goat-proof latch, or in any other area that goats can't reach. To keep it from getting moldy, make sure to put down pallets instead of storing it on the ground. Galvanized metal garbage cans with secure lids make the best storage facilities for other feeds. Rats chew through plastic and wood, which makes these materials less useful.

Some people have found a non-working chest freezer (which you can often get from an appliance store), with an added hasp and lock, a good option for storing grain and Chaffhaye.

Despite securing feed, rats, mice, and sometimes other animals, such as skunks, move into your barn, and into feeding and food storage areas. Any grain left in a bowl or feeder or dropped on the floor will entice them.

TECHNICAL STUFF

Rats dig holes, chew on wood, plastic, and even electrical wires, and they can spread diseases such as leptospirosis and salmonella. They leave their droppings everywhere they go. Four kinds of rats can plague the goat barn — Norway rat, roof rat, cotton rat, and woodrat. All of them are destructive and dirty.

You can choose to live with rats or fight them. I do both — I live with them until their digging gets out of hand or until they become so brazen as to run across the room in front of me.

You have several options for getting rid of rats:

WARNING

- **Barn cats:** Cats make good mouse and rat traps. Keep a couple of them in the barn and feed — but don't overfeed — them and they will work for you. Older cats are best because they're bigger and more experienced. Kittens can spread toxoplasmosis to goats, so make sure that you neuter or spay your barn cats.
- **Live trap:** This is the most humane method and one that I use most often. The Havahart XSmall works for rats and mice. You can buy one at a livestock supply store, a hardware store, or even Amazon.com.

  Set the trap with some bait; peanut butter works best. When the rat goes for the bait it steps on a part that causes the door to close. When you get to the barn in the morning, a fearful or annoyed rat is waiting in the cage. You then take it farther out into the country away from human civilization, let the rat out, and start over again.
- **Snap trap:** Look for the large snapping rat trap (although I've inadvertently caught a rat with a mouse trap).

  Set the trap with some peanut butter. When a rat springs the trap by eating the peanut butter, the trap kills the rat. Bury or burn the dead rat and reset the trap.
- **Rat poison:** You can purchase One Bite rat poison at any farm store. Break off a piece of the poison and put it in a rat hole where goats, dogs, and cats cannot get access to it. Watch for dying rats over the next few days. When you find a dead one, bury or burn it, and put out more bait.

  WARNING

  This poison is lethal to other animals that eat the dead rats. According to the Cascade Raptor Center in Eugene, Oregon, the most common reason that birds are brought in to them is because they ate an animal that was poisoned with rat poison and they're now sick.
- **Garbage can:** We smelled decaying animals in the barn for several days before my friend found four dead rats in the bottom of an open garbage can that had a little dog food in it. After rats jumped in, they couldn't get out. See the upcoming sidebar "A low-tech rat trap" for another option.

A LOW-TECH RAT TRAP

Rats are scourges to agriculture throughout the world. They destroy many crops and spread disease. A lot of different rat traps have been designed to combat them. Here's how to make one like the farmers in Zimbabwe use:

1. Bury a five-gallon bucket in the ground near a known rat hole leading to your barn or goat shed.
2. Cut off the ends of a corn cob so that it's about 6 inches long.
3. Push a 40-inch long, thick wire through the middle of the corn cob and widen the hole so the cob can spin freely.
4. Center the cob over the bucket; bend the wire down on each side at the top of the bucket and push the ends firmly into the ground on either side of the bucket.
5. Put 6 inches of water in the bucket and coat the corn cob with peanut butter.
6. Look for drowned rats, and recoat the cob daily.

» Considering your goats' body condition

» Feeding for special conditions

» Buying feeding equipment

» Building a hay feeder

» Building a mineral block holder

» Offering your goat some feeding options

» Planting a hedgerow

» Giving your goats some fodder

Chapter 6

Dinner Time: What and How to Feed Your Goats

Before you get your goats, you need to make sure you have all the feeding equipment they need, as well as a stockpile of food. All goats need certain kinds of feed — such as a mineral mix — but they have differing nutritional requirements depending on their gender and the stage of life they're in.

In this chapter I cover the feed-related equipment you need to buy or make, what food your goats need for adequate nutrition, feed alternatives, and how to store the feed that you need to have on hand.

Goats Don't Eat Tin Cans: What and How to Feed

Goats are ruminants and have a different digestive process from humans. (Chapter 2 explains that process.) Goats also are different from cattle and sheep, which mainly graze on grassy pasture. They are browsers, like deer, which means they prefer trees, bushes, and woody weeds; rather than standing still and eating grass down to the roots, they like to stay on the move, eating a bit of this and a bit of that (but not tin cans). Goats can learn to graze a pasture, but don't expect it to be "mowed." The grass helps supplement the goats' diet, but low grazing also can spread parasites.

Goats have specific nutritional needs, only some of which are met by the plants on your farm that they browse on. You have to provide feed for the needs that can't be met by browsing.

REMEMBER

Unless you have a lot of property with a variety of browse, feed will be your biggest expense in raising goats. Don't scrimp on goat feed — it will pay dividends in good health, milk production, and lower veterinarian bills.

In this section I give you a general idea of what and how often your goats need to eat for optimal health and production and suggest some special treats that will endear you to them.

Understanding the two types of feed

Goat feeds can be divided into two types:

- **Roughages** are high in fiber and come from the green parts of plants. They include grasses and browse (leaves, twigs, and shoots) from plants such as shrubs or blackberry bushes. (Rosebushes, too, as any goat that escapes the pasture will let you know right away.) Hay and straw are roughages that have been cut and cured, making them ideal for storage and later use. *Chaffhaye* is a roughage in a bag, which has the added benefit of containing good bacteria that aid in digestion.

  Roughages constitute the bulk of a goat's diet. They are key to a goat's digestion and add bulk to the diet, but they are low in energy (calories).

- **Concentrates** are complementary to roughages and are usually referred to as *grain*. They are low in fiber but high in either energy or protein. Some are high in energy, others are high in protein, and even others are high in both. The concentrates most commonly fed to goats include corn, oats, barley, and soybeans.

Feeding hay and alfalfa

Hay is the main source of nutrients for goats in non-grazing seasons or all of the time if they don't have access to browse. Grass hay provides a moderate amount of protein and energy for the goat diet. Legume hays, such as clover and alfalfa, usually have more protein, vitamins, and minerals, particularly calcium, than grass hays. This varies depending on the maturity of the hay or alfalfa and the way that it's cured and stored.

Each goat needs about 2 to 4 pounds of hay each day, although some of this need can be met by available pasture or other forage. Make it available free choice throughout the day when pasture is unavailable or feed twice a day when goats are also browsing.

TIP

You can feed alfalfa (and some grass hays) in pellet form if you don't have storage or if you want to mix it with grain. The goats don't waste so much alfalfa when it's in pellets, and you can limit who gets it by combining it with their grain.

Using Chaffhaye instead of hay and alfalfa

Chaffhaye (`www.chaffhaye.com`) is forage in a bag and substitutes for hay. It has been used in Europe for many years, but it's still fairly new in the United States, and you can't find it in some areas.

To make Chaffhaye, producers cut early non-GMO alfalfa or grass, chop it, mist it with molasses, add the culture *Bacillus subtillis*, and vacuum-pack it into 50-pound bags. The treated hay naturally ferments in the bag, adding good bacteria that's easier for goats and other livestock to digest. It provides more energy, vitamins, and minerals than dried hay.

Goats need up to 2 pounds of Chaffhaye per 100 pounds of body weight when you feed it as an alternative to hay. The nutritional value of one 50-pound bag of Chaffhaye is equivalent to an 85- to 100-pound bale of good-quality hay. It's easier to store and transport, and each bag has a two-year shelf life.

Chaffhaye is a good option for urban farmers, who don't have the necessary storage for hay or alfalfa, or goat owners in areas where good hay and alfalfa are hard to find or overly expensive.

TIP

If you want to try using Chaffhaye instead of hay, I recommend that you start small, buying only a little and seeing whether your goats will eat it. Some goats love it, but others refuse to eat it.

Feeding grain

Grain or pelleted grain mixes add protein, vitamins, and minerals to a goat's diet. Some are formulated specifically for goats (Purina goat chow, for example). Grain options include the following:

- **Whole grain:** This is the whole, unprocessed grain seed head.
- **Pelleted grain:** A product made from grain or grain byproducts milled into small pieces and then made into pellets by adding a binding agent.

  My goats and I dislike the pelleted feeds — they're too dusty, aren't as good on the ruminant digestive system as the rolled grains, are more likely to cause goats to choke, and apparently don't taste as good, either.
- **Rolled grain:** Nutritionally identical to whole grain, rolled grain is simply rolled so that it's flat.
- **Texturized grain:** Similar to rolled grain, texturized feed mixes usually have other ingredients added to improve nutrition.

If you have horses or donkeys with your goats, avoid feed or minerals that contain Rumensin (to prevent coccidiosis). Rumensin is toxic, and it takes very little to poison these animals.

I prefer to make my own mix of goat feed, combining one bag of whole barley (which is high in calories), two bags of whole oats, half a bag of peas, one-quarter bag of black oil sunflower seeds, and one-quarter bag of alfalfa pellets (using 50-pound bags). In a pinch, I have used wet COB (high-protein corn, oats, and barley with molasses) because it often goes on sale.

If you can get enough other goat owners together, consider working with a nutritionist at the local feed mill to formulate your own grain mix. I haven't tried this because you usually have to buy three tons or more and I don't have that much storage.

If you have livestock guardian dogs, avoid feed that contains cottonseed, because it can kill dogs that eat it.

Whether you feed your goats grain and how much you feed them depends on their purpose, the stage of their lives, their sex, and other factors such as pregnancy and lactation. Here are a few guidelines:

- **Kids:** You can raise kids entirely on hay and pasture or feed them grain for all or part of their first year to maximize growth. This is an individual decision based on your herd goals.

If you plan to regularly feed grain to your kids, start them on an 18- to 20-percent protein grain and decrease that to 16 percent after they're weaned. Give them access to grain in a *creep feeder* (a feeding area that other, bigger goats can't get to) at all times. (You can read about creep feeders in Chapter 13.) Giving kids *free-choice grain,* which means that the grain is available for them to eat whenever they want, helps prevent bloat and enterotoxemia because they start gradually and maintain the same or only slightly increased levels of feed. (See Chapter 11 for more on feed-related diseases.) When you change the volume or type of grain, do so over several days to allow the kids to adjust to the change.

- **Wethers:** Wethers can be susceptible to urinary stones and don't normally need grain. (The exception is goats that you're feeding to be large, such as pack wethers or meat goat kids.) They do fine on a grain-free diet if you give them good-quality grass hay. Even alfalfa can be problematic because of the calcium, which causes a certain kind of stone. If you must give them treats, select some of the treats I talk about in the section "Treats and snacks."
- **Bucks:** I don't give grain to bucks except during *rut* (breeding season). They can get run-down and not eat enough when they're breeding, and a little grain gives them much-needed energy. Remember to start slow with the grain, to avoid bloat and diarrhea.

Pregnant and lactating doe and senior goats need grain for the energy it gives. Read more about diet for these groups of goats in the section "Feeding for Special Cases."

Following a feeding schedule

TIP

I feed my goats twice a day, 12 hours apart. At the morning feeding, I load them up with grass hay to eat free choice, along with pasture during the times of the year that it's growing.

In order to get goats to use your pasture (if you have one), you need to decrease the hay supply or they'll just stay at the feeder and munch all day.

I give my does alfalfa during the last two months of pregnancy up through the time they're done lactating. (See the upcoming section "Feeding for Special Cases" to find out more about giving a pregnant or lactating doe everything she needs.) In addition to the twice-a-day feeding, in the winter when the ground is bare or covered with snow, I check the hay in the afternoon and refill it if it is empty or low.

I don't creep-feed but sometimes give the kids some grain during the first two or three months of life because they're growing (and I need some bribery to get them into the pen for evening lockup so I can get their moms' milk in the morning). I start with a small amount in a bowl for all the kids to share and gradually increase the amount to avoid gut problems.

Choosing organic — or not

If you are a purist and want to make sure that your milk or meat don't have any artificial chemicals in them, in some parts of the country you can find organic grain and hay. You may have to put some effort into finding organic products, and you can expect to pay more. On the other hand, if you sell milk or meat, you can expect to make more money in the market if you use organic feed.

Minerals are a must

Goats need supplemental minerals, vitamins, and other nutrients in addition to those they get in their hay, grain, and browse. Fortunately, with the growing popularity of goats in the United States, more and better mineral mixes are available.

Vitamins and minerals are essential to keeping goats healthy, making sure they're growing well, and assisting in reproduction and the development of skin and bone. The amounts of minerals and vitamins that goats get from their food will vary depending on what hay, minerals, and browse they have access to. If you have a large acreage with a variety of weeds and brush, your goats get more natural nutrients than goats that are raised in a limited area just eating grass hay.

You can supplement your goats' browse and feed them essential minerals and vitamins by supplying them with free-choice loose minerals or a mineral block. You can find goat-specific minerals in most feed stores, or you can request a special order. Some minerals are formulated specifically for meat goats and others for dairy goats. Some have salt and some don't. Goats prefer minerals with salt, but if you have to get a salt-free mineral, supplement it with a salt block.

WARNING

Never buy a so-called "goat/sheep mineral" because it doesn't have enough copper for a goat's needs. The amount of copper that a goat needs can kill a sheep.

Expect your goats to consume a quarter-ounce to one ounce of mineral mix per day. If you can't find goat-specific minerals, you can use a cattle or horse mineral. Goats and cattle have similar needs for certain minerals, but because cattle eat so much more, goats may have a problem eating enough of a cattle mineral to meet their needs.

Supplemental feeds

With good hay and an adequate mineral block (and grain for breeding and lactating animals), your goats get by just fine. But you can also give them some of these supplemental feeds to make them even healthier or, in some cases, to save money. Many of them can be left out in mineral feeders or bowls so the goats can eat them whenever they choose.

Beet pulp

Beet pulp is a byproduct of processing sugar beets to make table sugar. Processors add molasses to the pulp to bind it and keep the dust down and make it into pellet form for livestock feed.

Beet pulp adds fiber, protein, and energy to a goat's diet and contains calcium and phosphorus. It comes in 50-pound bags and is cheaper than grain, but doesn't supply as much energy and so shouldn't be used as a substitute.

WARNING

Sugar beet farmers started using genetically modified (GMO) sugar beets in 2008. If that concerns you, you may not want to give your goats beet pulp.

WARNING

I use beet pulp pellets in the winter as a supplemental feed. Make sure to hydrate beet pellets or shreds in hot water — they're a choking hazard.

TIP

Start by giving your goats a small amount unless you want to be disappointed and have to dump it out. The first time in the year that I offer the beet pulp, my goats are suspicious and don't want to touch the pellets, but by the end of the season they are fighting to get to them and slurping them down.

Black oil sunflower seeds (BOSS)

Black oil sunflower seeds are a relatively inexpensive supplement to goats' grain. They contain vitamin E, zinc, iron, and selenium and also add fiber and fat to the diet. BOSS make the goats' coats shinier and increase the butterfat in their milk. I buy 50-pound bags and mix the seeds into my goats' grain. They eat them shell and all.

Kelp meal

Kelp meal is made from seaweed and is a good source of iodine, selenium, and other minerals. Used as a supplement, it helps protect goats from iodine deficiency. Kelp also improves dairy goats' production, increasing milk volume and butterfat and helping decrease mastitis.

Kelp meal is expensive, and I can't always afford it. When I can, I buy kelp in 40-pound bags and give it to my goats free choice. Because they love it and fight to get to it, I have a hard time keeping it stocked.

Baking soda

Many goat owners offer free-choice baking soda to their goats, which aids digestion by keeping the rumen pH-balanced. Because my goats waste it, I provide it sparingly.

If one of your goats has a digestive problem, offer baking soda. It helps balance the rumen pH. Baking soda is also one of the treatments for floppy kid syndrome (see Chapter 13).

Nutritional yeast

Nutritional yeast has been found in studies to improve milk production in cows and goats, as well as to increase dry matter intake. It's also known to aid in digestion. I give my goats free-choice nutritional yeast about once a week, because they love it so much that I would go broke feeding it daily!

Apple cider vinegar (ACV)

Some goat owners add unfiltered apple cider vinegar to their goats' water. Old wives' tales abound about how adding ACV to water will lead to the birth of more doe kids, but I don't think it's ever been scientifically proven. ACV is full of enzymes, minerals, and vitamins, so giving to your goats can't hurt.

Treats and snacks

A lot of people want to feed their goats grain because the goats like it so much. But, like candy, just because it tastes good doesn't mean it's good for you. You can find plenty of other nutritious snacks for goats.

My goats love unsalted peanuts in the shell as a treat. You can buy a large bag of peanuts in big-box stores. They also come in handy when children come to visit and want to attract the goats.

Corn chips are a good grain substitute for wethers, because the saltiness encourages them to drink water, which helps prevent urinary calculi. (See Chapter 11 to find out more about urinary calculi.)

Fruits are always a good treat for any animal. Goats love apples, watermelon, peaches, pears, grapes, bananas (peel and all, if organic), dried fruit such as raisins, cranberries, and most other fruits. Just make sure that the fruits aren't in pieces large enough to cause choking.

Vegetables are a nutritious addition to any diet. Goats love carrots with their tops attached, celery, pumpkins, squash, lettuce, spinach, and other greens.

Avoid members of the nightshade family such as potatoes and tomatoes, which contain alkaloids, and plants with oxalates, such as kale. These can be poisonous to goats.

You can also buy commercially produced goat and horse treats for your goats.

Using Body-Condition Scoring to Fine-Tune Feeding

Body-condition scoring is a way to determine whether a goat is in good health or will be a good producer based on its fleshiness. You can use it to determine whether you're feeding your goats enough or too much.

Like people, some goats are constitutionally fatter or thinner, but no goat should ever be at the high or low end of the spectrum. You will invariably have a goat that is an *easy keeper* — that is, no matter how little you try to feed her she stays fat. Other goats have the opposite problem.

You can look at a goat from every angle, but without physically handling it, you can't really know its physical condition. For example, fiber goats or bucks with long coats may look well-filled-out, but when you feel their bodies you find that they're very thin. I learned this the hard way when I had a buck die. He had looked like he was perfectly healthy, but after he died I discovered that he was skin and bones covered with a lot of hair.

To determine body condition, you have to feel certain key points of the goat's body. The ribs and spine are the best indicators of body condition because the bones protrude there. *Body condition scores* (BCS) range from 1 (very thin) to 5 (obese). The ideal body condition score is a 3, although goats will be at 2 during certain times of the year, such as during rut, when the bucks forget to eat and run themselves ragged, or after a doe kids.

You have to practice body condition scoring to get good at it. After all, if you haven't felt many goats, you won't know what goats are supposed to feel like when they're in optimal condition. So don't be shy, feel any goats that you're around and try to determine their body condition so you can use this tool in fine-tuning your feeding program. Here are the guidelines for scoring a goat's body condition:

- **BCS 1:** The goat is emaciated. The ribs, spine, and shoulder blades are sharp and pronounced. The space between each rib is quite visible because of lack of fat between them. The vertebrae are sharp and noticeable. The flanks are hollow and the sternum (above the front legs) has very little fat. The loin has no fat covering it.

- **BCS 2:** The backbone is well defined but has some fat covering. The ribs are visible, but without a sunken area between them, and they're hard to feel where they come off of the sternum. You can feel a fat pad under the sternum, which you can move a bit. The loin is evenly covered with a small layer of fat.
- **BCS 3:** The backbone is well-covered with fat, and the vertebrae don't feel sharp. You can feel a fat pad on the sternum that can't be moved much. The ribs are barely visible, but you can feel them. You can't feel the ribs where they come off the sternum without pressing hard and trying to find them. You can feel a smooth, even covering of fat over the loin.
- **BCS 4:** You can't see the separate vertebrae in the spine. The sternal fat pad above the goat's front legs is thick and can't be moved. You have to search for ribs and can't feel the bones as they come off the sternum. The loin is covered with thick fat.
- **BCS 5:** The goat is clearly obese. You can't feel any individual vertebra or ribs. You may even see dimpling and hanging fat on various parts of the body, particularly the sternal region. The loin area may be so thick with fat that it jiggles.

If you have only a few goats, they usually are in comparable body condition, so adjusting feed intake up or down will help them all to attain the right condition. Just adding some grain to the diet of thin goats or decreasing the grain, alfalfa, or treats for a fat herd is often all you need to do.

If you have one goat that's very thin and is having trouble getting enough food, try moving her to a separate pen for a little extra feeding each night. If you add grain, make sure to start slowly. If a goat is extremely thin, but seems to be eating all right, consider testing for a disease that causes wasting, such as Johne's or CAEV. (See Chapter 10 for more on testing.)

If you have a goat that is clearly obese, you probably find that that animal is first to the feed and getting more than his share. Move that goat to a separate pen during feeding time and limit his feed.

Feeding for Special Cases

You need an overall program for feeding your goats to keep them at maximal performance, but at times you need to make exceptions for certain goats or categories of goats. Pregnant goats, milking does, kids, and senior goats need special attention and modified diets. (I tell you about feeding senior goats in Chapter 14.)

Pregnancy

Pregnant does don't have increased nutritional needs until the last two months of gestation, when the kids do 70 percent of their growing. They also need additional water throughout pregnancy. A feeding program for pregnant goats includes:

- **Early pregnancy (first 3 months):** Feed does to maintain their body condition or to improve their body condition if they are thin. You can meet their nutritional requirements with good hay or pasture, or some added grain for thin does. Unless they're lactating, does don't need grain in early pregnancy. Do not overfeed. Overfeeding can lead to complications such as hypocalcemia and ketosis. (Chapter 12 tells you about these conditions.)
- **Throughout pregnancy:** Monitor and compensate for pregnant does' increased water consumption. Pregnant goats can drink up to four gallons a day. Monitor body condition and adjust feed accordingly.
- **Late pregnancy (last two months):** Does' nutritional requirements increase greatly during this time, because the unborn kids are growing rapidly. Start grain gradually (just a handful a day) until your does are eating up to a half-pound of grain a day (depending on the goat size and breed) or half to two-thirds of their normal milking ration by the time they kid, in addition to hay. Gradually replace their hay with alfalfa so they get the proper balance of calcium and phosphorus. Continue to monitor their body condition and adjust feed accordingly; does carrying multiple kids need even more calories and nutrients.

WARNING

Make sure not to overfeed grain during pregnancy to avoid the risk of having the kids grow so large that the doe has birthing difficulties.

Milking does

Milking does, or does that are nursing their kids, have higher nutritional needs than other goats. You will have started your pregnant goats on grain and gotten them used to eating a substantial amount of grain during the last two months of pregnancy. Continue this feeding, even increasing it to several pounds a day, according to the doe's body condition and the number of kids she is feeding or the amount of milk she is producing. Also, make sure that she is drinking plenty of fresh, clean water.

Kids

Kids have different needs than adult goats because they're growing. If you're raising meat goats, you want good weight gain and so should feed the kids grain. Check out the sections "Feeding grain" and "Following a feeding schedule" for specific kid-feeding plans.

Getting the Basic Supplies

As soon as you know you're really going to get some goats, you need to figure out what and how many you need of each of the basic feeding equipment items. I talk about minimum requirements for these supplies in this section.

Bowls

You need feed bowls for grain, unless you just have a few wethers, who shouldn't eat grain. Feed bowls come in all sizes, shapes, and materials. You can find nice plastic or metal wall-hanging feeders with single or multiple sections. You can make your own out of wood — or metal, if you have the necessary equipment and skills. I use 5-quart plastic *feed pans* (a feed pan has a flat bottom, which makes it less likely than a bowl to tip over) to feed kids and the large 3-gallon plastic feed pans to feed adults when I feed in groups. Empty mineral tubs work well for feed and water, if you have a number of goats to feed, too. If your minerals don't come in a tub, ask friends if they have a spare one.

I use a stainless steel bowl for feeding does that are being milked on the milk stand because I found one that fits. See Chapter 4 for more on milk stands. Some milk stands come with a bowl.

Buckets

Goats drink a lot of water and unless you have a pond or a nice, clean creek running through your property, you need buckets or a *stock tank* (a large watering tank made for livestock). I prefer buckets to a stock tank, which is a lot larger and doesn't have to be filled as often. I like the buckets because they're easier to clean, and goats are notorious for refusing to drink water if even a piece of hay is floating on it.

I don't know of any rule of thumb regarding how many buckets you need per goat. It depends on how often you clean and refill them. At my farm, the goats get new water two or three times a day. I recommend starting with at least one bucket for every two or three goats (unless you have a huge herd) and seeing how that works before investing in more.

TIP

Buy dark-colored buckets. Algae is less likely to grow in a dark-colored bucket.

When I got my first goats I bought five-gallon buckets, thinking that they would work better because they held more and would have to be refilled less frequently. That worked great for about five years and then (not letting on how old I am) I realized that my back and arm couldn't handle carrying a five-gallon bucket. So I switched to the two-gallon size. Now I can carry one in each hand, which is still heavy, but at least I'm in balance.

In my experience, plastic buckets have only about a five-year lifespan. Over time, they crack and break. So consider that when planning your budget.

For the first three years I had goats I didn't have a nice barn, much less indoor plumbing for the goats. So I had to carry water from the house to the barn. If you live in a cold climate, you need to keep two things in mind: Goats prefer hot water, and hoses freeze. If you don't have a hot-water supply in or near your barn, plan to carry buckets of water several times a day. Fortex rubber-polyethylene buckets, which can be found at Tractor Supply or another feed store, hold up under rough handling when frozen.

If you have electricity in your barn, you can buy a plug-in heated bucket in either two-gallon or five-gallon size, or a submersible heater that can be put right in the bucket. These are a must in a cold climate, unless you want to refill buckets throughout the day.

Water supply

Goats need a consistent supply of fresh, clean water to grow properly, stay healthy, and do their best in milk production. Pregnant and lactating does have the highest water needs, and when the weather is hot and dry, all goats need more water. A general rule of thumb is between a half-gallon to four gallons each day per goat.

Using a human-made water supply

You can get by without a nearby water supply, but if your barn is any distance from your house, you'll find that hauling water and equipment back and forth is a lot of work. Plan ahead, even if it means digging a water line when you put in a new goat shelter or barn.

Even better is a sink and faucet with hot water — a dream that came true for me about five years into raising goats. Now I can wash feed bowls and buckets without schlepping them to the house. And with dairy goats and the equipment they require, it's even more useful. I don't have to get quite so mad when they step in or kick over a bucket of milk because I can clean up and start over.

At a bare minimum you need a hose that reaches the barn or area where the goats hang out. Even with just that, if you don't want to be at the beck and call of goats or if you actually have to leave the farm once in a while, you can use a stock tank and buy a float valve to attach to the hose. When the water is low, the valve opens to allow enough water through to refill the tank. Rubbermaid makes an inexpensive one that livestock supply stores carry.

Using a natural water supply

Some people are fortunate enough to have a creek or a pond that goats can use for a water supply. Both are convenient but have the disadvantage of being more liable to become contaminated, either from urine, feces, and debris or from becoming stagnant. If you plan to use either of these sources, get the water tested first to determine whether it's safe for drinking. If you are on a city water source, talk to your water supplier about how to do this. If you have well water, you are responsible for its safety and can contact your local health department for information on getting it tested.

Mineral feeder

Supplemental minerals are a must for goats, so you need a mineral holder or feeder to put them in. Look around your local feed store and ask other goat owners what they use. First you need to decide whether you will provide loose minerals or a block.

Loose minerals come in a bag and can be poured into a feeder; a mineral block is minerals that are compressed into a hard block using molasses to hold them together. Mineral blocks withstand moisture better because they already contain a certain amount of moisture in the molasses.

With loose minerals you have a variety of choices for equipment. I tried feeding pans that attach to the fence with limited luck. My kids figured out pretty quickly that they could jump up and stand in them, which contaminated the minerals. Some of the goats also like to paw at them when eating.

A good alternative for feeding loose minerals is a polypropylene mineral feeder, which you can get from a livestock supply catalog or store. It is sturdy, and kids can't stand on it.

You can find plans on the Internet for making a mineral feeder out of PVC pipe. This is an inexpensive alternative to a purchasing a mineral feeder and works well, if you install it securely. Goats are really hard on anything attached to the wall. I tried these feeders, but my goats eventually knocked them down.

I use minerals in a block now because my goats like them better. You can buy a holder designed specifically for mineral blocks, which even has holes in the bottom to allow moisture to escape. I've used them, but they also have a fatal flaw — they have no cover and so are easily contaminated.

I hadn't seen any mineral block holders that had a cover, so I had my friend Bob design one for me. You can make one of your own with the instructions later in this chapter.

Hay feeder

Any goat person will tell you that goats are terrible hay wasters. New to goats, my friend Bob spent hours conceptualizing the perfect hay manger that would prevent this waste, only to discover within a week that the floor of the goat shed was inches deep in hay. I'm not convinced that the perfect hay manger or feeder exists.

I don't recommend a *keyhole feeder* — one that has a larger opening at the top where a goat can put his head in, and a narrower opening where his head rests when he drops it to eat. The goat is unable to pull his head out of the narrower opening, which prevents wasting hay but it can cause damage or death to a goat that gets bashed from the side by a more aggressive goat.

I've also tried a homemade wooden, V-shaped open-top hay manger. It was set off the ground on legs, with slats on top in a V-shape down the middle. Within an hour after I set it out, a goat jumped into the hay on top and got stuck. I recommend that if you get one of these feeders, you put a top on it. That will keep goats out and if you feed outdoors, a top will keep rain and snow off of the hay.

Now I use two different types of hay feeders — a sheet metal box frame feeder that I can move and hang on any fence, and a large round feeder made from cattle panel. (See instructions for making the round feeder in the section "Building a Simple Hay Feeder," later in this chapter.)

The box feeder is a rectangular box with slats going across the top of the front to the bottom of the back. When goats pull out hay, some of it drops into the bottom of the feeder, so not as much is wasted. The round feeder has openings at different levels all around so goats can stick their heads through to get hay.

I keep the large feeder in the barn where most of my goats live and smaller ones in the smaller areas where I keep bucks, wethers, boarders, or goats in quarantine.

You can buy a poly plastic version of the box frame feeder I use at some feed stores or online at `www.enasco.com`. Make sure not to buy the kind that has an open bottom or has the bars so widely spaced that hay can fall out.

You need at least one box feeder for every four goats. The round feeder can easily feed 8 to 12 goats at a time, depending on their size and temperaments. Often one or two goats drive others away from a feeding area, so I like to make hay available in at least two locations in each goat area. That way everyone gets to eat.

Storing feed

You can store hay anywhere that's dry, sheltered, and separate from your goats so you can dole it out. (If goats have free access to hay they will eat or spill it all.) A section of the barn or a storage building is best, although I had a friend with limited storage who kept a quantity of hay outside on pallets covered with a tarp.

Unless you live in a very dry climate, make sure to carefully check the hay before using it because it can get moldy when stored outside. Mold can cause sickness or even kill goats.

Chaffhaye is a little easier to store than hay because it is vacuum-packed so it can withstand the weather better. You can store it in a barn, a garage, or even outside on a pallet.

You need separate containers to store bags of grain, sunflower seeds, and other concentrated feeds such as beet pulp because they attract rats. The rats on my farm are so persistent that they even chew through plastic, so I use 30-gallon galvanized steel garbage cans. One can holds two 50-pound bags of feed. I loop a bungee cord through the handle on the lid and attach it to the side handles to ensure that it stays closed if a goat manages to get into the feed storage area. The best way to prevent goats from getting into feed is to store the feed in an area where the goats can't ever get to it.

Building a Simple Hay Feeder

The simple feeder in Figure 6-1 is easy to make, easy to use, and holds enough hay to feed 8 to 15 goats (depending on their size) for several days. You probably want to use it for only 10 goats or so, however, because they may start fighting when eating in such close proximity.

The only problems I've had with this hay feeder is that the goats sometimes tip it over when it's empty, and every so often three goats (kids, usually) stick their head through the same panel and get stuck when they try to pull back at the same time. You can attach the feeder to a wall or post to prevent goats from tipping it over, but doing so limits the number of goats that can feed from it because they don't have access to the part of the feeder that's against the wall.

You can make a 38-inch-diameter circular hay feeder from a cattle panel. If you want a larger one, use a longer section of cattle panel. Remember that the goats have to get their heads to the hay in the middle, so don't make it too wide.

FIGURE 6-1: A simple homemade hay feeder.

You need only a few tools and materials for this hay feeder:

- A 16-foot section of cattle panel
- Bolt cutters
- A metal file or a grinder
- Heavy wire, fasteners, or zip ties

Follow these steps to put together your hay feeder:

1. **Using your bolt cutters, cut a 10-foot length from the cattle panel.**

   Make sure to cut at the end of a section so that no metal ends protrude.

2. **File or grind any sharp parts.**
3. **With another person helping, roll up the section of panel until the two edges meet.**
4. **Secure the ends together with zip ties, strong wire, or metal fasteners.**
5. **Place the feeder in the desired area, fill it with hay, and watch your goats go for it!**

Building a Mineral Block Holder

You can buy a mineral block and put it in a bowl or a holder designed for mineral blocks, but you will soon be frustrated. Goat kids love to jump up on anything and everything, and mineral blocks are no exception. All goats will relieve themselves when the fancy strikes them — even if they're standing near or over a mineral block. This means that your goats' mineral block will soon become contaminated with urine and feces, directly and from little feet grinding into it.

My friend Bob came up with a really simple idea for a wooden mineral block holder that keeps the mineral block off the ground and also covers the top of it, so when those kids start jumping they just land on wood. You can buy the wood for this, or make it from pieces you have lying around. The inside dimensions (10½ inches) are important because that's the minimum required to hold a standard-size mineral block. (You can make it larger, if you want.) You need a solid post or wall to attach it to.

To make the mineral block holder you need the following equipment and materials:

- Hand saw or circular saw
- Drill
- Pencil
- Yardstick or other measure
- Carpenters' square
- Level
- One eight-foot long untreated 2 x 6 board
- One sheet of ¾-inch untreated plywood
- Fifty-six 3-inch deck screws
- Twelve 1½-inch flat-head screws

To make your own mineral block holder, follow these steps:

1. **Measure your 2 x 6 and mark with pencil at 12-inch intervals.**

   Use the carpenters' square to make sure the ends are even. (***Note:*** You can make the top section from a 2 x 4, but you need to use a 2 x 6 for the bottom so the mineral block can rest on it.)

2. **With your saw, cut eight 12-inch pieces from your 2 x 6 for the sides.**
3. **Measure your plywood.**
   - Mark two 13½-inch by 13½-inch pieces for the top and bottom.
   - Mark two 6½-inch by 16-inch pieces for the top and bottom braces.
4. **Cut all the pieces of plywood.**
5. **Using eight screws (two for each corner), screw together four of the 12-inch 2 x 6 pieces to make a 13½-inch square.**

   To make the box square, make sure each piece of wood is attached on the inside on one end and the outside on the other.
6. **Attach the first piece of plywood to the top of the square with 12 screws.**
7. **Assemble the bottom section by repeating Steps 5 and 6.**
8. **Using a level to ensure that it is even, attach the top and bottom sections to the post or wall with eight screws each so the bottom of the top and the top of the bottom are 16 inches apart.**

   Attach the bottom piece 14 inches from the ground or higher. The plywood should be on top for the top piece and on the bottom for the bottom piece. Don't situate it too high or too low for your goats to comfortably eat the mineral.
9. **Place the top brace with one end inside the top section against the center of the front piece of the 2 x 6 and the other against the post at a 45-degree angle and attach with three screws on each end.**
10. **Place the bottom brace with one end under the front 2 x 6 piece of the bottom section and one end against the post at a 45-degree angle and attach with three screws on each end.**

    To make a better fit, you can cut the end of each brace in a 45-degree angle.
11. **Set the mineral block snugly into the bottom section.**

Figure 6-2 shows you a completed mineral block holder.

FIGURE 6-2: You can build a simple mineral block holder.

Providing Supplemental Feeding Options

If you really want to spoil your goats and provide them with more variety, you have a number of options. You can grow them a garden, create a hedgerow of plants for them to eat, try a fodder system, or give them access to brush they normally can't get to.

Growing a goat garden

Goats and gardens are symbiotic. Goats supply the fertilizer, and the garden supplies the fruits and vegetables. I use the mix of straw and manure that I get from mucking out my barn to supply my garden with the compost it needs for healthy soil. It only seems right that the goats should get something in return.

You don't have to dedicate the whole garden to your goats (unless you want to). Just make sure that whoever the garden is for, you keep it away from the fence line. Goats have no discretion about when to eat a plant and don't think about waiting for it to grow to maturity or bear fruit.

Here are some plants you can grow with goats in mind:

- **Sugar snap (or snow) peas:** Sugar snap peas go in the ground as early as February or March and are usually ready to eat in June. Both peas and their stems and leaves make great goat treats. They're high in fiber and contain lots of vitamins and minerals.
- **Carrots:** Carrots are a goat favorite, full of fiber, vitamins (especially vitamin A) and minerals. Even if you want to use some of the carrots yourself, your goats will be more than happy to eat the green tops for you. You'll want to cut carrots up to avoid a choking hazard.
- **Beets:** Beets are another vegetable that goats don't waste — they enjoy the root and the greens. Also packed with vitamins, they have one downside: They're high in oxalates. Too much can lead to urinary calculi (usually in wethers) and toxicity that can cause muscle tremors, weakness, and other problems.
- **Sunflowers:** Sunflowers are a goat favorite. They can eat the whole plant, as well as the seeds. Make sure to put some netting over the flowers, or the birds will eat them first! If you have enough, you can break off the heads and give it to your goats for a treat and some vitamins A, B, D, and E.
- **Squash and pumpkins:** Squash and pumpkins can add some protein, zinc, and vitamin E to your goats' diet. Because they keep well throughout the winter, you can grow a bunch and dole them out over several months. The best way to serve them is by cutting or breaking into smaller pieces. And the seeds have a mild dewormer effect, according to one study.
- **Greens:** Greens of all types provide diversity to a goat's diet. This includes lettuce, spinach, kale, parsley, and more! Greens add dietary fiber, but watch out for feeding too much of the oxalate-containing greens such as spinach, swiss chard, and kale.
- **Berries:** Berries are like candy to goats. Where I live, the land supplies plenty of invasive blackberries, but I also grow strawberries and raspberries. The hard part for me is having to share with the goats.

» **Tree fruits:** Tree fruits such as apples and pears require less time and energy. They provide lots of fiber, minerals, and vitamins. However, both apples and pears contain a high amount of fructose, which can lead to digestive problems. So limit how many you give at one time, cut them up to prevent choking, and remove apple seeds — which contain cyanide.

» **Melons,** including watermelon and cantaloupe, are excellent treats for goats, especially in the heat of the summer because they contain so much water. The whole fruit, as well as the vines, is safe and nutritious.

TIP

Make sure to start slow when you introduce a new food to your goats. Although the fruits and vegetables listed here are safe, they shouldn't make up a large part of the diet. And you may have that one aberrant goat who's allergic to something.

Feeding with garden plant waste

Whether you grow specifically for your goats or just for yourself, there is more you can do with the plants in your garden. When it's time to remove the old plants, don't forget the goats! They can eat much of the waste, as long as it isn't moldy or bug-ridden.

For example, when your peas or beans are done producing, tear down all the old vines and throw them over the fence to the goats. They also like cornstalks, even without corn, as well as sunflower stalks and many other plants. When you do your annual cutting back of raspberry and blackberry vines, give them to your goats!

WARNING

Avoid giving plants in the nightshade family, such as potato and tomato, and limit the amount of garden waste from oxalate-containing plants you give your goats at any one time.

Storing garden produce

Root vegetables and tree fruits such as apples and pears can last into the spring, if you have enough and you store them properly. This will allow you to dole them out over the winter, providing added nutrients when the pasture and other plants aren't growing so much.

Another option for providing food throughout the season is to dry it. Many fruits and vegetables can be dried and used in the winter.

Raspberry leaves are traditionally given to pregnant animals at the end of their pregnancy to help stimulate good contractions. Consider drying some if you have goats who will be kidding. And blackberry leaves will do in a pinch.

Creating a Hedgerow

The concept of a hedgerow is centuries old. A hedgerow is much like a hedge, except it's made up of several plants of varying heights, while a hedge is traditionally composed of only one. It's usually at least 10 feet wide and at least 20 feet long, although the size will depend on usable space available. It can also provide added benefits such as dividing up the pasture or creating a windbreak, depending on where you put it.

Choosing a location

A hedgerow can take up to four years to become firmly established, so you need to choose an area your goats can't reach while it's growing. In addition, you'll probably want to have your hedgerow on one side of a fence, in case the goats "accidentally" eat through it and go where they don't belong.

Think about what your goals are for this planting. If it's only to feed goats and you want it to last a long time, putting a fence between the plants and the goats will prevent them from eating your plants until they've grown enough to be reached. Alternatively, you could put it in an area that you only allow your goats to enter occasionally — which will enable you to control their browsing.

Determining what to plant

The first thing to think about is what native plants that goats can eat are in your area. These most likely will be the most successful.

You also need to consider the sun exposure in the area you're planting. This will help guide you regarding the plants to choose.

Choose a planting day that isn't too hot. The best times for starting a hedgerow project is fall or spring — when you would normally plant perennials.

Start with a row of taller plants, such as fruit or other trees. Good choices are apple, pear, or hawthorn. Avoid cherry, peach, plum, or other plants with seeds that contain cyanide. Other goat-friendly choices include willow, bamboo, fir, or lilac.

TIP

If deer frequent the area, make sure to put some wire fencing around the trees until they're established.

Underneath the trees, you can plant woody shrubs that goats are so fond of. These can include roses, cane fruits such as raspberries or blackberries, blueberries, honeysuckle, and salal. Use your imagination, and check first to see whether goats can eat them.

Plant them in front of and in the area between the trees, and then fill in front of that with the third grouping.

In a third grouping, growing under the trees and shrubs, you can include ground plants or herbs, such as dill, parsley, oregano, plantain, clover, or grasses.

After planting, make sure to water and mulch everything well.

Keeping it surviving and thriving

To make your hedgerow successful, you need to start right. The first step is to get rid of weeds and remove any woody plants. Then cover it with cardboard and mulch thickly and thoroughly. This is where your goat manure comes in handy. Cover the cardboard with composted manure and then cover that with 3 or 4 inches of good soil. Wait at least two weeks, and you're ready to plant.

To keep the hedgerow growing and healthy, water regularly and keep up with weeding. If you find that you're getting deer damage, rethink how to protect your plants.

You'll eventually have a natural fence that will last through the ages.

Supplementing with Fodder

Fodder systems are becoming more popular as a way to feed livestock these days. Such a system can be particularly useful in areas that suffer from drought or are in the desert and have fewer plants available.

A fodder system allows a goat owner to sprout a grain such as barley or oats in plastic trays. One benefit is that the feed produced is less expensive than traditional feed (hay). It also has the advantage or substituting for forage in the event of drought conditions.

A disadvantage is the time required to create the feed. The seed must first be soaked for at least 12 hours; then put it in trays and water up to three times a day for a week. Although the watering can be automated, doing so creates an added expense to the cost of the trays and lighting. Figure 6-3 shows a fodder system used for sprouting supplemental grains.

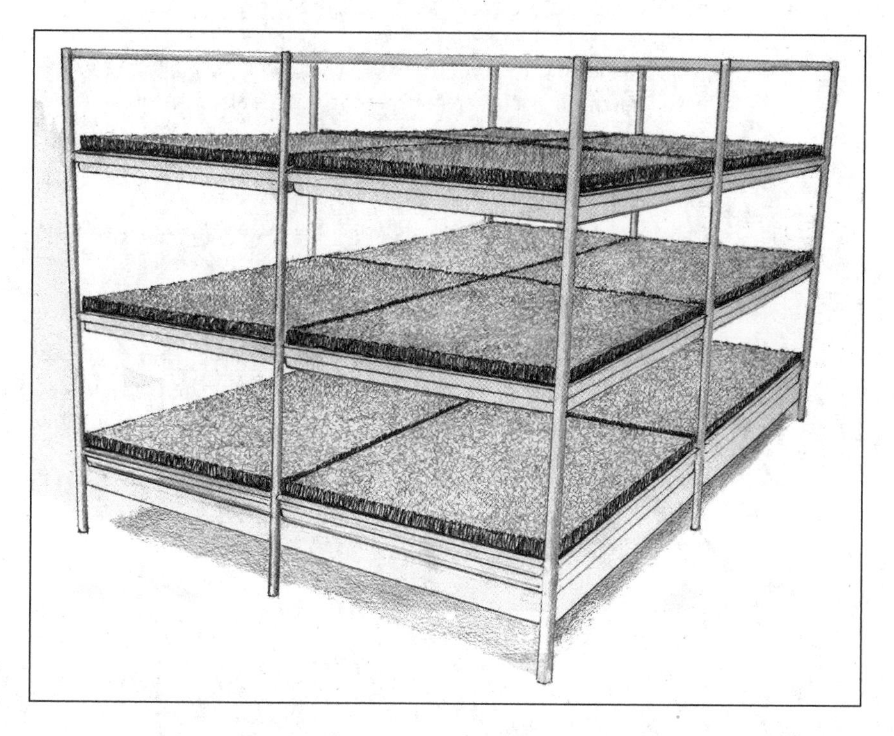

FIGURE 6-3: A fodder system.

According to one study, barley and oats provided less non-fiber carbohydrates, meaning that they had less starch than other types of grain. Barley fodder also had the most calcium and was found to increase milk production in cows.

Farmtek (`www.farmtek.com`) offers systems of varying sizes. Large systems can supply all the feed, while smaller ones will easily supplement your goats.

CUTTING BRUSH FOR YOUR GOATS

Most of us need to keep our goats restricted to a limited area, both to protect them and to protect the plants outside of that area. But that doesn't mean that they can never have some of those plants. You can bring them to your goats.

When got my first two goats, I lived on a property that was at the bottom of a hillside. Most mornings I would let the goats out and take them on a slow, supervised walk up the hill, where they ate mouthfuls of whatever struck their fancy. They loved it, we all got some exercise, and they were able to supplement their pasture and hay diet with the woody plants they were designed to eat.

I can't do that now for a couple of reasons: I have too many goats to control and they would have to walk past areas that have bushes and plants I don't want them to eat (RIP butterfly bush).

When I moved to my current property, the fenced area had not just grass in the pasture, but quite a few of the shrubs and other plants they like to browse on. But those are long gone and they're left with leaning over the fence to the neighbor's property and pruning the apple trees, blackberries, and scotch broom.

Now I bring the brush to them. Every so often I walk out into the woods with my clippers and a box and load it up with fir branches, salal, blackberries, wild roses, and any other good stuff I can find. They love it, and it keeps their rumens healthy!

» Knowing which goats are right for you

» Using a sales contract

» Getting them home safely

Chapter 7
Getting Your Goats: Choosing, Buying, and Bringing Goats Home

I knew before I got goats that I wanted a dairy breed, and I always assumed it would be Nubian because I liked the way they look. Then I spent two years after I bought my farm learning everything I could about goats and finally decided to buy two Nigerian Dwarves.

After several false starts — from driving through pouring rain unable to find the farm where I hoped to get my first goats to missing out on the goats I'd planned to get because someone else bought them first — I got lucky and bought two does from a breeder who went on to mentor me for many years. (My partner refused to turn around when we had trouble finding *that* house!)

You might get lucky and find the perfect goats on your first attempt. But if you don't, just keep trying. This chapter takes you through the important points to consider when selecting goats, finding them, picking healthy ones, and closing the sale.

Choosing the Right Goats for Your Needs

Nothing is worse than buying some goats, bringing them home, getting attached to them, and then discovering that they won't work out in your situation, are the wrong type, or aren't what the seller represented. In this section, I talk about some characteristics to consider when looking for goats to buy.

Goats need company

A lot of people make the mistake of getting only one goat. It's a mistake because goats are *herd animals* — they live and move in groups and respond to each other's cues — so they need each other. One goat does not a herd make.

I got two does for my first two goats because I knew I wanted to milk and I could handle two goats and a few kids. If you're getting a larger breed or want to start slow with pets or brush eaters and, maybe, eventually start milking, consider getting a doe and a wether — sister and brother, because they know each other well. Which brings me to another goat characteristic: They get bonded to their family members.

I've heard stories of goats that were separated for years and then recognized each other when reunited. I sold two goats from the same herd to a farm, where they were initially kept in different areas. When they were together again, one followed the other around and cried when they were apart, even though they were different breeds and hadn't been particularly good friends before I sold them.

REMEMBER

You'll have a much easier time if you buy a pair from the same herd and, unless you want them to be unrelated for breeding purposes, from the same family. They adjust more quickly if they're with someone they know.

Function matters

Many people like the idea of getting goats but are unsure whether they want to get in deeper at a later date — for example, by milking, shearing, or eating them. If you love cheese (who doesn't?), it may be a good idea to start with dairy goats; if you like knitting, cashmere goats may be the best choice.

Believe it or not, goats actually prefer to hang out with others of their own breed, so even if you're getting a couple of wethers to begin with, you should consider starting with the breed that can perform the function you dream of.

Size matters

Unless you have unlimited space and feed, you need to consider the size of the goats you get. If you live in the city or have a small acreage, consider small breeds such as the miniature dairy breeds for milk, the kinder (a Nubian/Pygmy cross) for meat, or the Pygora (Pygmy/Angora) for fiber. These may be the only types of goats legally allowed in a city.

On the flip side, bigger is better in the evolution of some breeds. If you are raising Nubians or some of the meat goats, you want large or well-muscled — but not fat — goats. (See Chapter 6 to find out about body-condition scoring.) Likewise, if you're planning to backpack with goats, go with a larger breed. You can learn more about ideal size of the breed you are considering from a *breed club* (an association focused on a specific breed of goat).

Horns can hurt

Even if you object to removing horns, think carefully before you bring home a goat with horns. (Chapter 9 gives you the pros and cons of *disbudding*, or removing horns.) Horns can do more than cause physical pain to another goat or person; they can also hurt your pocketbook when you have to replace a fence, pay for a lawsuit, or pay medical or veterinary bills.

Find out whether the breed you want is one that is normally kept horned. Fiber goats, for example, need horns because the horns provide them with natural temperature control in the heat. Some meat goats are also not normally disbudded.

If you get a breed that is normally disbudded or *polled* (naturally hornless), keep in mind that if you plan to breed them and sell kids, they may not be as easy to sell if they still have horns. You also won't want to keep horned and hornless goats together because they aren't evenly matched and the hornless goats will lose any battle. Horns, especially on miniature goats, are often at eye height, and a careless toss of their head at a fly or other annoyance can injure your eye. You may also get tired of removing your goats from the fence repeatedly.

Registered or unregistered

A *registered goat* is one that meets the standards of appearance for its breed and is recorded in the herdbook of the goat association for that particular breed. A registered goat usually is a purebred but may be a crossbreed (called an *American* or an *Experimental*.) See Chapter 3 for more about crossbreeds.

If you just want a couple of goats to love, spend time with, and use for help keeping down the noxious weeds, then you don't really need registered goats. Registered goats cost a little more, but they aren't necessarily any better.

Registered goats have some advantages over unregistered goats. They are required to have identification such as a tattoo or a microchip, which can be helpful if they are stolen. Registration also gives you some assurance that the goat is the breed and has the potential the seller claims. Registered goats are more valuable, so if you want to sell surplus kids you can usually get more money for them than for unregistered ones. And most goat shows require entrants to be registered.

TIP

If you're buying registered dairy goats, ask to see relevant milk records generated by the Dairy Herd Improvement Registry (DHIR), if the seller is on *milk test*. Owners of goats on milk test send milk samples, along with information on amount milked by weight, to the DHIR monthly for recordkeeping and testing. These records will let you know about production by a specific goat or her dam or granddams over time. The information will give you an idea of whether the goat you're buying has any likelihood of living up to claims a seller has made regarding milk production. (I tell you more about DHIR in Chapter 15.)

Most registries also have a complaint process, so if you believe that a seller has acted unethically and contradictory to registry requirements, you have somewhere to turn besides the courts.

Looks count: The basics of conformation

In goatspeak, *conformation* refers to correctness of the body structure as compared to the ideal standard. Judges consider conformation when judging goats in a competitive show and appraisers look at it to determine how well a goat represents the breed. Knowing a bit about criteria for the ideal goat of the type you want helps when you're selecting goats for your herd.

Goats potentially live longer, are healthier, and are more productive if their structure and udder attachments are correct. They kid more easily if their rear ends and back structures are sound, milk more if their body capacity and width are good, and so on. These criteria are basically the same for all types of goats, with an emphasis on the udder for dairy goats, muscling and length of loin for meat goats, and type and fineness of fleece for fiber goats.

You'd need a whole book to cover the variations in conformation standards among different types of goats (and such books have been written). Appraisers use the following characteristics to determine whether a goat has good conformation:

- Masculinity for bucks and femininity for does
- Set and correctness of legs
- Shape of head
- Length of body
- Udder attachment
- Muscling (in meat goats)
- Level *topline* (the line down the back of the spine)
- Shape of scrotum
- Width of chest
- Width of *escutcheon* (area between back legs)

Goat registries certify individuals (usually judges) to appraise dairy goats — giving them a score based on their conformation and how they compare to written standards. Before you purchase goats, ask the sellers whether they have had their goats appraised and if so, whether they will share that information with you.

TIP

If you are interested in finding out more so you can more knowledgeably select goats for your herd, I recommend that you read further on conformation for the specific breed you want. *Dairy Goat Judging Techniques* by Harvey Considine and George W. Trimberger (Dairy Goat Journal) is *the* book on dairy goat conformation, but it is out of print and hard to find. You can get information on conformation for other types of goats online.

Finding Sources for Goats

Finding goats isn't hard. When you start looking for them, you'll see them everywhere. I recommend that you first try to find goats close to home. Doing so enables you to visit the farm and see the goats before you agree to purchase any.

If you're looking for a rare breed or have very specific needs, finding the goats you want locally might be a little harder, but take your time and use some of the resources I talk about in this section. After you get a lead, call or email that person and ask about his goats. If you're still interested after that, plan a visit to his farm to see his herd and the goats he's selling.

Visiting local feed stores

You'll be spending a lot of time at your local feed store getting the supplies you need for your goats-to-be and then continuing to buy feed and supplies after your herd is established. Get acquainted with the proprietors. Although the bigger chain stores are frequently cheaper and carry more products, mom-and-pop feed stores are a great resource. They are usually willing to fill special orders or help you find an elusive product that you're having trouble locating.

Often the owners or clerks at the feed stores know the farmers in the area who shop there and are willing to direct customers to them. (They may even *be* the farmers you're looking for.) They also usually have a bulletin board with livestock and farm-related ads, business cards, and flyers — a good place to start in your search for goats.

Reading the agriculture paper or thrifties

I've had mixed luck with the regional agriculture paper and the thrifties or nickel ads. I've found goats advertised in them, but I've also had very few inquiries when I advertised there.

Nevertheless, these publications are a good place to start looking for goats for sale in your area or a short drive away. The agriculture newspapers have the added advantage of writing about livestock and may even feature a breeder with just the goats you are looking for.

Checking out Craigslist

Craigslist, an online service for buying and selling, jobs, and personal connections, is a good resource for finding goats locally and often at a competitive price. I use it regularly and always get responses, even if I don't always sell the goats I'm advertising.

To use Craigslist, you just go to the site for your city or a close city and look in the farm and garden sale section. Craigslist is also a good place to find farm equipment and fencing that people are selling for a good price.

Surfing breeders' websites

You can find people who breed the type of goat you want by going directly to their websites. Just search for the type of goat you want and the word *breeder*. You'll find a number of websites, usually with pictures, pedigree information, and sales pages. Some breeders also include information on their management practices and helpful hints for raising goats.

Unless you want to pay for shipping, narrow your selections to nearby breeders. If the breeders you find aren't local, ask them about reputable breeders or people they've sold goats to in your region of the country or for other resources.

Joining registries and goat clubs

If you're sure you want to buy a specific breed, join the breed club or the registry for that breed. A breed club promotes the breed they're interested in and sometimes sponsors shows. A registry is usually a nonprofit association that keeps a herdbook on goats. When you join, you get a member handbook or contact information for other members who have the type of goat you want.

Joining a registry has some other advantages:

- You pay less to register your goats. Most registries have a fairly high price for nonmembers to register and a more reasonable one for members.
- Registries have written standards that you can use to evaluate goats that you might buy.
- They may host shows, special sales, linear appraisal sessions, herd testing, or other events.
- They often have a complaint process if you're cheated or victimized by a seller. They don't take action on your behalf but do provide notice of unethical sellers.

Even if you don't plan to register goats, these organizations are a good way to get in touch with breeders. Like a lot of breeders, I sell some of my goats unregistered for a lower price.

TIP

You may also be able to find a local or regional goat club to join. These clubs often offer a newsletter and sponsor shows. They may also hold conferences to educate goat people and give them a chance to network. When you talk to people selling goats that you're interested in, ask them about what clubs or goat-related organizations they belong to.

Going where goat people congregate

Your next county or state fair is likely to have a goat exhibit, which is a really good way to find out more about goats, look over some goats, and network with other goat people. Goat exhibitors use the fair as a place to acquaint people with their animals and to market them.

In regions with a large goat population — Texas comes to mind — goat shows happen quite frequently. Read the paper, watch the feed store bulletin board ads, ingratiate yourself with goat owners, and find out when a show will occur.

Some areas also have an annual goat day or goat conference. For example, the Northwest Oregon Dairy Goat Association (NWODGA) has an educational goat conference every February. Few goats are sold there, but people network, offer goat products for sale, have a raffle, and often hand out sales brochures. It's a good venue to meet goat people and find out more about goats that they or others are selling.

Making Sure You Get a Healthy Goat

Just as you wouldn't want to buy a car with a bad transmission or a dog that you know has hip dysplasia, you want to make sure that goats you purchase are healthy and aren't going to rack up a lot of vet bills (unless you're interested in rescuing mistreated or sick animals). In this section I share a number of techniques that help you determine whether a goat is in good health.

Asking questions

You can find out a lot just by asking questions. Then you can further protect yourself by making sure that the claims made about the goats are in a sales contract. (The upcoming section "Protecting Yourself with a Contract" gives you the goods on getting a contract.)

Ask the following questions to help determine whether a goat is healthy:

- What diseases do you test for? What kind of results have you had with testing? (See Chapter 10 for more about testing.)
- Do any of your goats have a transmissible disease? How do you handle that? (See Chapter 11 for more about common diseases.)
- What is your feeding program for newborns? (See Chapter 13 for more about newborn feeding.)
- What vaccinations do you do? (Chapter 10 tells you more about vaccinations.)
- Have you had any goats die from an undiagnosed illness in the past few years? What happened?
- Have you had a history of abortion in your herd? Explain.
- For meat goats: What kinds of market weights do you get for your goats?

- For fiber goats: How much fiber do you get on average from your goats and what type and quality is it? (See Chapter 18 for more about types of fiber.)
- For dairy goats: Are you on milk test? How much milk do you get from the goats or their dam or buck's dam? Do you have DHIR records for the doe or her relatives?
- What do you feed your goats, including minerals? (Chapter 6 tells you more about feeding goats.)
- Will you give me the names of three people you have previously sold goats to?

If the goats you plan to purchase are located too far away to visit, you won't be able to examine them or the herd they're coming from. Besides getting answers to the previous questions, you can take a couple more steps before agreeing to purchase them:

- Ask specific questions about characteristics that you might find on examination. For example, "Does this goat have any defects or has it had any illnesses?"
- Ask for pictures of the goat from different angles.
- Ask for copies of any health records on the goat.

Examining the goats

If you live close enough or can afford to travel, go to the seller's farm to see the goats. Ask to see the goats that you're interested in purchasing (or if they aren't born yet, to see their dams) and any health records the seller has. You can use this opportunity to check not only for sickness but for quality. Spend some time on the farm, evaluating the goats and talking to the seller about feeding and other management practices.

Look at the goat from a distance, observing how it moves and whether it limps or favors any leg. When you get to the goat, check its body:

- Evaluate its weight. You need to put your hands on the goat to determine whether it's bony, fat, or average weight. (See Chapter 6 for more about body-condition scoring.)
- Check the body for any lumps, swellings, or other abnormalities.
- Look for extra, split, or micro-teats.
- Notice whether the coat is dull, dandruffy, or missing patches.
- Check the eyes and nose for crustiness or mucus.
- Look for signs of diarrhea.

- Pull down the lip and check the gums for anemia. (The gums should be pink.)
- If the goat is lactating, inspect her udder for lumps, disproportion, or pendulousness. Ask to milk her if you're purchasing her for milking.

Observing the home herd

Observe the rest of the herd while you are at a farm to get an idea of how the goats look overall. (This will also give you an opportunity to see how the owner has set up her goat area and barn and get some ideas.)

Look for the same things from a distance that you're checking on individual goats you're interested in. And make sure that

- The herd looks well-nourished.
- The goats are ruminating.
- The pens are passably clean.
- Sick goats are isolated from well goats.

Protecting Yourself with a Contract

The days of handshake agreements aren't entirely gone, but they're on their way out. Never assume that the other party to a goat sale hears what you're saying. People usually hear what they want to hear, whether good or bad. As a result, proving an oral contract can be difficult, if not impossible.

I've bought and sold goats without contracts over the years, and I have been burned only once. But once is more than enough. To make sure that both the seller and the buyer are agreeing to the same things, you need a contract. In my experience, most goat sellers don't understand the importance or necessity of contracts.

Putting together a basic contract is likely to fall to you, and doing so is important for making sure you get what you want and what the seller says you're getting.

One case where you won't be able to make a detailed contract is when you buy goats at auction. If you buy a goat at a livestock auction, don't expect any assurance that that goat meets even minimum expectations. People take their unwanted animals to auctions, so avoid those goats if you want to make sure to get quality, health, or any other guarantee.

A contract is a legally binding agreement that creates an obligation to do or not do a particular thing. It puts in writing the agreements made by a buyer and a seller, helps prevent misunderstandings, and protects both parties in the event that they have a later disagreement or conflict. A written sales contract needs to include at least

- The effective date of the contract
- The names and addresses of the buyer and the seller
- A description of the goat, including
  - Name, if applicable
  - Breed
  - Gender
  - Whether the goat is registered
  - Whether the goat is for breeding or not
- Payment terms, including
  - Price
  - Deposit to be paid
  - Date(s) for other payments
- Health certificate and testing information, including
  - Party responsible for health certificate, if required
  - Tests already done and results
  - Party responsible for the testing, if not yet done
  - Outcome if tests are positive, for example, exchange/replacement of goat or refund
- Outcome if goat is infertile, if sold as breeding animal
- Action to be taken if doe sold as bred is not pregnant
- When seller's responsibility for change in goat's condition ends and buyer's starts — usually when transport begins or goat changes hands

Ask breeders or check the Internet for goat sales contracts. Some breeders include them on their websites. You can use one that has already been written and make the necessary modifications for your situation.

Bringing Your Goats Home

The way you move your goats from their old to new home depends on where you buy them. If you purchase your goats from a breeder who lives across the country, the breeder is responsible for getting the goats to a shipper, and all you have to do is find out where to pick them up. You normally are responsible for the cost of shipping and a health certificate, even when the buyer does the legwork. If you bought the goats from a neighbor and they are trained to lead, getting them home will be simple — just put them on leashes and lead them home. If you're buying goats in any other situation (and you probably are), you need to figure out in advance how to get them home.

In this section I discuss how to transport your new goats, the effects of transportation on goats, and what to do when you get them home.

Transporting your goats

You can transport goats in a variety of ways, depending on their size and how many you're moving at one time. If you can, avoid putting goats in a large area where they can more easily fall or get injured. Also, if you won't be moving your goats in a temperature-controlled environment, avoid transporting in extremely hot or cold weather.

Some of these methods won't work for large goats, but you can transport goats in

- Pet carriers or crates with straw or wood shavings for bedding. If you're moving the goat in an open truck bed in cold weather, cover the crate with a rug or tarp to keep the wind down and keep the goat warmer.
- The back of an SUV, van, or the back seat of a car. I cover the seat with a tarp and towels for those invariable "accidents" that occur during transport.
- On the towel-covered lap of a passenger. Goats that are being held and aren't standing up will not pee on you, but they will poop.
- The back of a truck with a canopy. Make sure to put down plenty of straw.
- A horse trailer or another trailer with fencing or cattle panels to make it high enough to prevent escape. Cover an open trailer in extreme weather conditions to protect the goats from rain and wind.

REMEMBER

Regardless of how you transport your new goats, to make the trip as stress-free as possible, do the following:

- Load the goats carefully.
- Make sure they have adequate bedding or padding.
- Start, stop, and take turns or curves slowly and smoothly.
- If your trip will take many hours or days, provide the goats with hay during the trip and stop every 3 to 4 hours to let them eat, drink, and regain their equilibrium. If you want to turn some heads, you can even take a lead-trained goat out for a walk at a rest area. ("What kind of dog is that?")

Quarantining new goats

If you're getting your first goats, you don't have anyone to quarantine your new goats from. You just get them situated in their new digs. But if you're adding goats to an existing herd, you need to quarantine the new goats for at least 30 days, if at all possible. This means that you need an area with adequate shelter that completely separates your herd from the new goats. Quarantine protects the other goats from any unknown or undisclosed health problems that the new goats might have.

During the goats' quarantine time, do the following:

- Have them tested for CAEV or any other diseases you want to test for, unless the seller has provided you with documentation that the goat has been tested and had negative results. (See Chapter 10 for more on routine testing.)
- Observe the goats for signs of any disease such as soremouth lesions or abscesses. (See Chapter 11 to find out about common ailments.)
- Watch how the goats adjust to your feeding and management program.
- Do a fecal analysis and deworm if necessary. If you're unable to do a fecal analysis, routinely deworm all goats in quarantine.

Watching for signs of stress

Even when you start with healthy goats, transporting can stress them emotionally and physically. Emotional stresses include

- Leaving their mothers and friends
- Losing their standing in the herd and having to establish a new position
- Being in unfamiliar surroundings

Physical stresses can include

- Being moved to a transport vehicle
- Prolonged standing in a moving vehicle
- Temperature extremes, rain, and wind
- Lack of exercise
- Insufficient food and water intake
- Crowding or being moved with unfamiliar goats
- Being bullied by more aggressive goats

At best, the stress of shipping only causes a goat to have a depressed appetite and not seem quite herself, but she snaps out of it in a few hours or days. Remember, she has to adjust to a new environment away from the security of everything she has ever known.

Blood tests show that a goat needs about three hours after being transported to stop having a physical stress response, but the move's effect on the goat's immune system can last longer.

WARNING

At its worst, the stress of transport brings on what is known as *shipping fever* — causing pneumonia and sometimes diarrhea. Signs to look for include temperature of over 103.5°F, nasal discharge, coughing, rapid breathing, or rattling in the chest. Contact a veterinarian if your new goat has any of these signs.

To minimize the effects of transport stress, give the goat plenty of water (warm or hot if the weather is cold and spiked with molasses if she isn't drinking), goat Nutri-drench, and some probiotics, and watch him closely.

Watch for bullying that seems excessive or dangerous, as they redetermine their status in the herd or among the new goats, and separate the bullies.

TIP

Bullies can be handled in a couple of ways. You can separate them and then intermittently reintroduce them to the herd to see whether they can behave. Or you can monitor the behavior, and spray the bullies with a garden hose or a spray bottle when they misbehave.

Eventually you can expect the new goats to settle in to their surroundings and be back to their normal selves.

» Giving goats basic and advanced training

» Finding and keeping a good helper

» Keeping your goat active

Chapter 8
Working with Your Goats

When you get a fix on goat dynamics and start working with and training your goats, you'll be amazed at how easily they learn. If you're new to goats, though, you might need some time to figure out what to expect from them. (They need to adjust to you, too.)

In this chapter, I tell you which goat behaviors are normal, how to work with and train goats in basic and more advanced tasks, and how you make sure they have proper supervision. I also give you fun ways to keep your goats active.

Identifying Normal Goat Behavior

I love studying the social aspects and idiosyncrasies of other animals. (People could learn a lot from them about how and how not to get along.) Human and goat dynamics actually have a few things in common.

Goat herds have a set structure, with leaders and followers. Some goats are decision-makers, and others go along with the decision. Sometimes goats find interesting ways to exercise power.

I had a goat with kidding problems: Some years she couldn't get bred and in others, her kids died during or shortly after birth. She loved kids, though, and developed a habit of staying with all of them while the moms and other goats went out to browse each afternoon. She was the only "auntie" goat I've ever met. Other goats carry their urge to nurture a little further, even going so far as to steal a kid from another goat. And many a kid, especially one with two or more siblings, will take a suck from any teat in sight.

Besides the behaviors I discuss in this section, goats have a habit of checking things out with their mouth. (This is probably where they got the reputation for eating everything.) Never put health records or registration papers near a goat if you want to keep them. I've even had goats chew up ribbons they won at the fair.

Establishing a "pecking" order

WARNING

Like chicken flocks (and our society for that matter), goat herds are hierarchical. Might rules, but so does nepotism. Sometimes you may feel inclined to intervene in your goats' bad behavior, but don't do it unless someone is getting hurt. Goats set the rules, and you can't change them without damaging the goats.

Mostly, you won't see your goats intervening in bullying either, lest they become the target. I have known a couple of does that walked between two fighting does, as if mediating the dispute. This behavior usually occurred before a fight really got going.

Bow to the queen

Every goat herd has a dominant female, known as the *herd queen.* She usually leads the way and decides when to go out to pasture. She gets the best sleeping spot, the primo spot in front of the feeder and, if she is a dairy goat, she gets to be milked first.

If another goat tries to change things, beware! The herd queen won't like it and will let the other goat know by her behavior. (See the section "Biting, butting, and mounting.")

The herd queen's kids are royalty by birth. The herd queen lets them share in the best eating spot next to her. She will defend them if any other goats try to get them out of the way. Unless they are challenging her for position, the only time other goats seriously take her on is if she is bullying their kids.

The herd queen has responsibilities, too. She tests new plants to determine whether they're edible and she also stands off predators. When a new dog comes on my property, the herd queen is often in the front of the herd, snorting and threatening to butt. My original herd queen, Jinx, once charged at and butted a cat she didn't think belonged in the barn.

BOYS WILL BE BOYS

When I was in high school, we had a hallway that was referred to as senior hall. Boys, mostly seniors, would line the hallway and give a hard time to girls who had to pass through the hall to get to class. I dreaded going down that hall, not knowing whether I was going to be insulted or even groped. I had a flashback to that time one day when I was observing my goats.

I watched from my house as a teenage doe escaped through the fence and trotted down the road toward an area where she could find blackberries, apples, and other goodies. As she approached the fence line that separated the bucks from the does, she slowed down and then stopped. Ahead of her were four heads sticking through the fence. The adult bucks were lined up, snorting, slobbering, and making all kinds of noise to attract (or sexually harass) her. She decided to chance it and rushed by, as far away from them as she could get. Senior hall all over again!

The herd queen usually retains her position until she dies or until she becomes old and infirm and another doe fights and wins the position.

The head buck leads

The *head buck* is the counterpart to the herd queen. He is usually the biggest and strongest (and often the oldest) buck. In my herd, the head buck is a Nigerian Dwarf who is smaller than his Oberian herdmates. He took over the position when the prior head buck died. He had been second in the hierarchy and the Oberians — who were destined to become bigger — were young and smaller than him, so he was able to maintain his position.

In the wild, the head buck takes up the rear of the herd, defending against any danger coming from behind.

Bucks also fight for the top position but, like the herd queen, a buck retains his position as head buck until he dies or a younger, more dominant buck challenges him and wins when the head buck gets old.

Biting, butting, and mounting

Goats sometimes communicate by biting. Some don't bite at all and others bite a lot. For the first few days after I sold a milking doe, Daisy, to another farm, the owner called me, distraught. She reported that Daisy was biting all of the other goats. She said she couldn't keep her because she was afraid her goats would get hurt and would start biting. I put her off each time, and within the first week the biting had stopped. Daisy was just trying to establish her place in the herd — near the top — and it worked. After that, she had no reason to bite.

Like biting, butting serves a role in the goat world. Goats butt to bully others out of their way, to establish their place in the herd, as a form of play, or to fight, often during rut.

Butting is one reason that keeping horned and dehorned goats together is unwise. The dehorned goats are at a distinct disadvantage and can be seriously injured. Unlike sheep, who run at and ram each other, goats normally rear up on their back legs and come down to butt.

Goat kids start mounting each other even when they're only a few days old. They are practicing to be grownup goats, but they're also attempting to establish dominance. As they get a little older, the mounting takes on a sexual connotation.

Goats, both male and female, respond to sex hormones by mounting each other. Mounting behavior tells you when a goat is in heat, especially if you have a wether in the herd. You can use him as an indicator of a doe in heat, just like a *buck rag* (a rag rubbed on a stinky buck, which a doe in heat responds to).

Establishing a place in the herd

The most common reason for butting and biting is to establish a place (as high as possible) in the herd.

When I let one of my first two goats out of her kidding pen for the first time, the other doe immediately started fighting with her. They butted heads until they were bloodied. I had to separate them and then try again, because the first time Kea was attempting to take over as herd queen. The second time I let Jinx out, she and Kea butted heads a few times and then Kea stopped. Jinx had maintained her position.

When you introduce a new goat to the herd, the lower-status goats are usually the first to fight. They want to maintain or raise their position in the herd, and the newbie is a threat.

Playing

Goat kids often butt each other, but usually gently. They use butting as a form of play, as they learn how to be goats. The same is true of mounting behavior. They are playing at being grownups. They are also trying to establish their role in the hierarchy.

WARNING

Sometimes you see goats just rear up and come down, rather than hit heads. They don't hurt each other, so don't try to stop the behavior when it is directed at another goat. On the other hand, if they exhibit this play behavior with you, *do not allow it.* A goat, especially a buckling, who is allowed to butt humans as a kid will grow into that mean goat you always hear about that's unsafe to be around.

Fighting

Goats fight over food and during rut. Goats, especially those with higher status in the herd, butt or even bite others when they perceive competition for food or, if they're bucks, for the opposite sex.

WARNING

Watch bucks during rut to ensure that they're safe from each other and to separate them when they aren't. I had a buck, Black Bear, whose rear leg was broken by a relentless and aggressive buck during breeding season. (He got around on three legs after that and even managed to climb a fence and breed a few of the does in his impaired state.) You also need to make sure not to turn your back on a buck during this time because they also can be aggressive toward humans.

Conducting Basic Training with Your Goats

You don't want goats that live up to the stereotype of stinking, butting, and being generally untrustworthy. And routine care is a lot easier if your goats have some basic training and are used to being handled. Even if you don't want show goats, you can get more enjoyment from goats that don't run every time they see you.

In this section I tell you about teaching your goats some basic manners, give you tips for handling goats, and show you how to use a collar to control your goats.

Collars are not just decorations

Collars are a useful tool for handling goats. They're necessary for showing a goat or getting it from one place to another, like in and out of a vehicle for transport. A collar is essential for restraining a goat when grooming, unless you have a *milk stand* (a piece of equipment that a goat stands on with her head secured), also sometimes called a *stanchion.*

A collar can be anything from a piece of baling twine to a plastic chain link collar to a fancy silver choke chain, depending on the situation and your budget. If you have only a few goats, you can buy collars at the local pet supply store. If you have a larger herd, consider purchasing from a pet or goat supply catalog, which gives discounts for bulk orders.

Measure your goats' necks before buying collars so that you get the right sizes. Standard collar sizes are

- 22 to 24 inches for an adult doe
- 26 inches or more for a buck
- 14 inches for a kid

For first-time goat owners I recommend the plastic jumbo chain-link collars sold by goat supply catalogs and some hardware stores. These come in 20- or 40-inch lengths that you can size to fit each goat. Goats can wear these collars all the time, because if a goat gets his plastic collar caught on something, the links break apart.

If you have a breed such as Oberhasli, in which all the goats look similar, you can also color-code them to help identify goats, especially new kids. One year, I got the mothers and kids matching collars, with a different-colored attachment link for each of the kids. It allowed me to identify them easily from a distance.

As you get more comfortable with your goats, consider using a regular dog collar when you need to control a goat and letting it go collarless the rest of the time.

Goats that wear collars all the time are at risk of choking if the collar gets hooked on something.

Handling goats regularly

Goats need regular handling or they can get wild. A wild goat will run from you, struggle when you handle it, and try to get away at any chance. This kind of behavior presents a problem when you need to groom it or do routine maintenance and care, such as hoof trimming or vaccinating. Nothing is worse than having to chase down and capture a sick goat or having to drag and lift it into a vehicle for a vet visit.

On the other hand, a goat that is handled regularly knows you are in charge and is more likely to come when called or at least not hurt your back or arm trying to get away.

A goat's initial comfort with being handled depends on its age, its temperament, whether it was bottle-fed, and how much it was handled before you got it. The more you handle a goat, the easier it is for both of you — the goat won't fight and you won't risk getting hurt. Plying the goat with a treat such as peanuts doesn't hurt, either.

Here are some important things to remember when you start handling a goat:

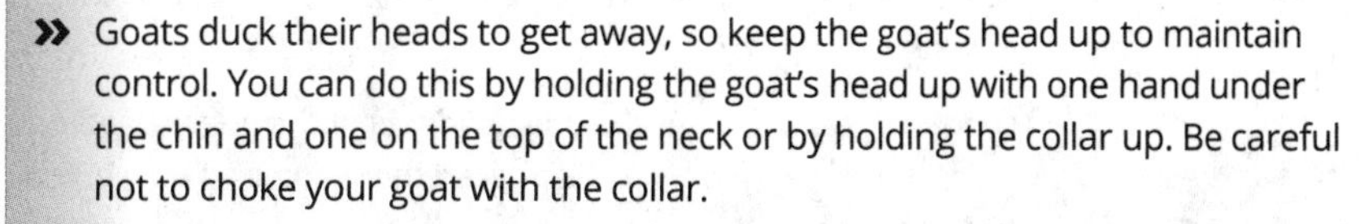

- Goats duck their heads to get away, so keep the goat's head up to maintain control. You can do this by holding the goat's head up with one hand under the chin and one on the top of the neck or by holding the collar up. Be careful not to choke your goat with the collar.

WARNING

- To catch a fleeing goat, grab the back leg. Catching a goat by a front leg may break the leg.
- Avoid chasing a goat. You can stress him out, leading to other problems. Try luring him with food instead.
- To handle a horned goat, firmly grasp the base of the horns to lead her.
- If you expect a veterinarian visit, catch the goat in advance. Restrain the goat in a pen or by tying her to a fence.
- Use treats to lure a goat that is resistant to handling. Peanuts, carrots, or apple chunks are good choices.

Walking goats on a lead

Lead training is essential if you're planning to show your goats. A jumping, fighting, or recalcitrant goat doesn't fare well in the show ring (see Chapter 17 for more about showing goats). Even if you don't show goats, teaching them how to walk on a lead makes them easier to manage.

When you lead train, try to recruit a helper to gently push from behind when the goat stops walking. To lead train, take the following steps every day for at least 10 minutes:

1. **Put a collar on the goat.**

   Make sure it isn't too tight or so loose that it slips over the goat's head.

2. **Attach a lead and position the collar.**

   The collar should fit at the top of the goat's jaw area and behind the top of the head.

3. **Walk forward a few steps, pulling slightly on the lead.**

   If the goat follows, continue walking. Stop every few steps and reward the goat with a small treat or praise. Then start again. Gradually increase the distance the goat has to go to get the treat.

WARNING

   Do not drag the goat or pull too hard on the leash because it can block the windpipe and cause the goat to collapse. The goat will drop to his knees, sometimes exhibiting what appears to be a seizure, but will recover quickly.

4. **Stop after a few steps if the goat stops, and then try again.**

   Use a treat such as peanuts, apple chunks, or corn chips to encourage the goat to walk.

   Lead the goat; don't let the goat lead you. If the goat is leading well but tries to go past you, say "Stop" or "Get back," then stop and turn in the opposite direction. When the goat is stopped by the rope, wait a second, and then start to lead in the direction you turned.

5. **When the lesson is complete, lead the goat back to the herd and remove the collar and lead.**

After a goat is more comfortable walking on a lead, you can practice less frequently. Use this method to get the goat to the stanchion for routine care or just take her on an occasional walk on a lead.

Teaching basic manners

Goats need basic manners so they don't hurt you or someone else, especially if they have horns. Jumping up is one bad goat habit and butting is another. When I had my first kids, I let the buckling, Harry Potter, jump onto my back and shoulders when he was little. He was so cute! Then one day I was out in the barn in a tank top when suddenly I felt sharp hooves and a weight on my back. Now that he was bigger, it wasn't so cute. I broke him of the habit by sitting with my back to a wall when I went in the barn, so he had no target to jump on. Needless to say, I never let another goat jump on me.

My daughter inadvertently taught the same buckling that jumped on my shoulders to butt people when she played with him by pushing on his head. Years later, when I went into the buck pen during rut, Harry Potter reared up and came down right on my hip bone. I avoided serious injury, but I kept my distance after that. Now I tell visitors not to push on goats' heads, even in fun.

For mannerly goats, follow these tips and share them with any visitors to your goat herd:

- Never push on a goat's head. Pushing simulates butting and teaches the goat that butting humans is all right.
- Never let a goat kid jump on you or anyone else. It won't be a small kid forever, and those hooves hurt!
- Don't let a goat stand with its front legs on you. The goat will eventually get bigger and can knock you down with those front legs.

» Lead-train your goat for basic handling. Doing so lets your goat know who's in charge.

» Don't let children chase your goats. Being chased makes the goats more fearful and potentially harder to handle; it can also make them sick from stress.

Do not let children ride the goats. In addition to making them more fearful, being ridden can break their backs. Goats are not built to carry people.

Moving Up to Advanced Goat Training

Goats are very smart and, despite their independence can be taught to do much more than follow on a lead or come when called. Start training when goats are kids if you want to succeed with advanced training. Older goats can sometimes learn tricks or to pack, but they're more resistant than younger ones.

Training success depends on your spending a lot of time with the goat and being consistent with your teaching.

In this section I tell you how to teach your goats tricks with a clicker, how to housebreak goats, and how to teach them to pack or to pull a cart.

Teaching tricks with a clicker

You can teach a goat to do almost anything with *clicker training*, a positive reinforcement technique that was developed to train dogs. (You can learn more about this method at `www.clickertraining.com`.)

You need a clicker, which is a mechanical device that makes a click sound, and treats such as peanuts or flakes of cereal. By combining the click with a treat, you reinforce that the goat is doing the right thing. You need to start by getting the goat to make a connection between the clicker and a treat. To do this, click the clicker and then give the goat a treat about 20 to 30 times. Your goat begins to associate the clicker with food and eventually responds to just the clicker so you don't have to supply a treat every time.

After you've shown the goat that treats are tied to clicks, you can start training. You train by issuing a command ("Come," for example), and then clicking as the goat does what you want it to do and giving the goat a treat after he completes his task.

Of course, the goat won't follow your command without practice. If the goat doesn't respond to the command or does the wrong thing, you can just say "wrong" or another word, and then try again.

Always click as the goat does the behavior and then give the treat. If you give the treat first, you risk the goat being caught up in eating and not noticing the clicker. Your goal is to have your goat do a certain behavior without getting a treat. You can also use this method by clicking when your goat does a normal behavior that you want to reinforce — for example, pawing at the ground.

TIP

If you're trying to teach something more complicated, such as "play dead," break the task into smaller incremental steps, and teach each one. The goat quickly learns to associate the action with the click sound and with the treat.

With patience and a lot of treats, you can teach a goat to carry out some surprising tasks. In 1998, Mary Ellen Nicholas wrote in *Ruminations: The Nigerian Dwarf Goat Journal* about her goat Bentley. She trained him to roll over, come, fetch, close a door, play dead, wave, pick things up, and "pray" (go down on his front knees) using clicker training. She even taught him to pee before going into a building.

Using an obstacle course

Goat obstacle courses are popular at goat shows and state fairs. Some 4-H groups use them as another way to get children involved with their goats. You can train a goat to use an obstacle course with a clicker or just with treats and repetition.

Use your imagination to set up a goat obstacle course. You can use hay bales, hula hoops, a 2 x 4 held up by two cinder blocks, old tires, or whatever else you can think of to get a goat to climb up and over, go around, or jump. If you set up a ladder that isn't too steep, you can also train a goat to climb that.

Housebreaking

Most people would be horrified with the thought of bringing livestock into their homes. But baby goats are so cute and doglike in their intelligence that, once you know them, you can't resist. And they can learn not to pee in the house, or where they shouldn't. Pooping is another story, but those goat berries are dry and easy to clean up.

I know that you can housebreak a goat because I did it accidentally quite a few times. I've occasionally raised kids in the house for the early part of their lives. (I was also lucky enough to have a job that allowed me to bring them to work in a crate that I kept next to my desk. Thanks, Jan!) When any of these goats would stretch out (boy) or crouch (girl) to pee, I whisked them up and out the door to the deck. I didn't have a clicker and didn't even know about them, but the goats learned not to pee in the house.

I had a friend who knowingly housebroke his goat kid. He put down lids from photocopy boxes with newspaper in them. When Malakai started to pee, he moved him to the box. Within a few days, the kid was using the boxes on his own.

I don't recommend raising goats in the house. I think they need to be with other goats and outside so they can get used to the climate. And they love to nibble on everything, especially paper. (If you want this book to stay in one piece, don't leave it on the table near a house goat.) If you need to raise a goat in the house or want to bring your goat in the house from time to time, use the methods above or try clicker-training to housebreak your goat.

Preparing goats for packing

You can train any kind of goat for packing, but you're better off looking for a goat with certain qualities. You best choices for pack goats are

- **Large.** Dairy wethers are the most highly valued goats for packing. The bigger they are, the more they can carry.
- **Friendly and energetic.** Pack goats have to work with you as a team, and so they'd better like humans. They also need to be able to take on the task of walking and carrying supplies.
- **Intelligent and curious.** Intelligence and curiosity are good indicators that the goat can be successfully trained to pack. They will be in new and different situations where curiosity, rather than fearfulness, is important.

TIP

If you're serious about packing, join a pack goat organization such as the North American Packgoat Association (www.napga.org) or an Internet discussion group so you can learn from others and get more resources.

To train a goat to pack, first make sure that he is accustomed to being handled and is calm. The basic steps for pack training are

1. **Teach your goat to accept being tied.**

   This is important because the goat will need to be tied when you stop to camp, eat, or just rest.

   Make sure your goat has a sturdy collar that isn't too loose. Tie the goat to a gate or fence at back height with a rope or lead that is a foot to a foot-and-a-half long for a short period of time.

Stay nearby to ensure that your goat doesn't get hurt. If the goat starts to get tangled up, calmly untangle him and tie him to the fence again. The goat should accept being tied up after only a few sessions.

2. **Teach your goat to follow you.**

   You will be leading your goat when you pack. Follow the steps in the section "Walking goats on a lead," earlier in this chapter.

3. **Teach your goat to stand.**

   Your goat will need to know to stand when you're putting the pack saddle on him, or at other times on the trail.

   When you are lead training and you come to a halt, say "Stand" or "Stop." Pull up on the lead rope if your goat doesn't stop. Practice this repeatedly, rewarding your goat when he complies, until he gets it.

4. **Teach your goat to wear a pack saddle.**

   He will need a pack saddle with a *pannier* (a bag to carry gear). Otherwise your goat will just be hiking, rather than packing.

   First, show your goat the pack saddle and let him examine it. Gently place the pad and saddle on his back. Tighten the *cinch strap* (the strap that goes around the goat's body), then fasten the *breast collar* (the strap that goes in front of the goat's chest) and, last, fasten the *rump strap* (the strap that goes around his rump). Check to make sure that two fingers fit between the goat and the cinch strap. Figure 8-1 shows a goat ready for packing.

TIP

   Give your goat several opportunities to get used to wearing the pack saddle alone or with a pannier that is empty and to go through Steps 1 through 3. Take him hiking with it empty. Before you load up the pannier to go on a hike, determine how much weight your goat can safely carry. A rule of thumb is about 20 percent of body weight. Never overload a pack, which can injure the goat.

When your goat is comfortable walking or hiking with a pack saddle and pannier, you can add the supplies you need and plan a packing trip. If you're in a pack goat club, you can start with a group outing.

Pack goats can participate in community service activities. In 2009, *Goat Tracks* magazine reported on a volunteer project where goats and their owners went into the Modoc National Forest in California to clean up the remains of an isolated pot farm that had been busted. They hiked in to a steep, brushy area and packed out hundreds of feet of hoses, propane tanks, bottles, pots and pans, and pounds of trash.

FIGURE 8-1: A pack goat can carry up to 20 percent of its body weight in a pannier.

Training goats to pull a cart

A goat can't pull a large load because it isn't built like a horse. Generally, a goat can pull about twice his weight, which includes the equipment — harness and cart. You first need to teach a goat to pull an empty cart before moving on to pulling a cart with someone in it. Take plenty of time with training, and train the goat away from other goats and distractions so you can both focus and avoid problems.

Goats that pull a cart are called *harness goats.* As with pack goats, wethers are the best goats to train to pull a cart. They get bigger than does, aren't at the mercy of their breeding cycles, and don't stink like bucks. The ideal harness goat is energetic, intelligent, and friendly. He isn't afraid of strange objects and doesn't whine.

TIP

Before you train a goat to pull a cart, make sure that he is accustomed to being handled and is calm.

Train him on a *bridle* (the head harness used to steer the goat), using these steps:

1. **Let him sniff and mouth the bridle.**
2. **Put the bridle on his head, tightening all straps.**

   Praise the goat and give him a treat.

3. **Take him for a walk while he's wearing the bridle.**

   Do this every day for a week so he can get used to wearing the bridle.

4. **After he's gotten used to the bridle, attach the reins.**

   Take him for a walk and hold the reins so he can get used to them. Now he is ready for both.

After your goat is used to the bridle, get him used to wearing a *harness* (the equipment that the goat wears to pull the wagon). The first time you put the harness on him, put his collar and bridle on first. To train him to use a harness, take the following steps:

1. **Let him sniff and mouth the harness so he gets used to it.**
2. **Put the harness on him.**

   Tighten the cinch strap, and then fasten the breast strap, and finally fasten the rump strap. Praise the goat and give him a treat.

3. **Take him on a walk, using the collar to attach the lead.**

   Give your goat several opportunities to get used to wearing the harness.

4. **After he has gotten used to the harness, attach the lead to the harness instead of the collar.**

   Intermittently, gently pull on the harness to simulate the feeling of a cart being pulled. Praise the goat and give a treat.

5. **Take your goat on a walk while he is wearing the harness and bridle with reins, and teach him to stop, go, and turn to the left and right.**

TIP

   When you want him to stop, say "Whoa" or "Stop." Pull the reins if your goat doesn't stop. Say "Go" or "Walk" when you want him to go and "Left" to go left and "Right" to go right. Practice these commands every day for a week or so, rewarding your goat when he complies, until he gets it. He may have trouble going in front of you if he has walked by your side in the past. Bringing a helper to walk by his side with a leash attached to the goat's collar will make the process easier.

6. **Practice with a *travois* (a simple triangular wood frame).**

   Before you get a real cart, you need to simulate a cart in a safer way. Attach a travois to the harness and take your goat through the commands. To make a travois, get two long poles and a shorter pole and lash them together into a triangular shape. A travois normally has a platform or netting between the poles, but you don't need one if you're just getting a goat used to pulling a load. Practice with a travois for a week or so before attaching your goat to a cart or wagon.

7. **Attach your goat to a cart.**

   You can buy a cart from Hoegger Goat Supply (`www.hoeggergoatsupply.com`) or American Cart and Harness (`www.k9carting.com`). Hoegger Goat Supply also sells shafts to make a regular wagon into one a goat can pull. Let your goat look at and smell the cart. When he loses interest in it, hook him up and go (see Figure 8-2).

FIGURE 8-2: Put your goat to work by teaching him to pull a cart.

Supervising Your Herd

Goats don't need a lot of supervision during the day. Mostly they just need a safe place to spend their time and exercise, clean water, and plenty of food.

In this section, I tell you about how to meet your goats' needs and find a good helper or helpers to care for your goats when you aren't available or need a vacation.

Meeting their social needs

Goats are social animals and like to spend time with other goats and with their humans. Goats need other goats to keep them company and meet their social needs. If you think that you or your dog can take the place of another goat, think again. They need one of their own kind to be around all the time.

And sure, after you have two goats, you can leave them to their own devices and check them just twice a day during feeding and water if they're in a safe area. But remember that if you spend time with them, you'll be calmer (goats have a calming effect, which I tell you about in Chapter 18) and they will be even happier. And you'll end up with better trained goats that really like humans.

Evaluating the time you have to be home

Before you get goats, consider whether you have the time and can be locked into the daily schedule of goat care. Many goat owners have full-time jobs and can't be home during the day, except on weekends, and yet they do fine with their goats. If you're in this situation, you need to schedule time to care for your goats before and after work. I've had to do this at certain times in my life. As long as your goats are safe and you leave them plenty of hay and fresh water during the day, they'll be all right.

REMEMBER

Plan to spend a minimum of a half hour in the morning and a half hour at night on routine goat chores. If your herd is very large, or if you have to milk, expect to spend more time. Routine goat chores include feeding, changing water, cleaning buckets, and watching your goats to make sure they're healthy.

On weekends or when your schedule allows you need to make time for routine care (see Chapter 9). If you have only a couple of goats, an hour on a weekend is adequate for hoof trimming and brushing. Even if they don't need their hooves trimmed, you'll want to check them out and hang out with them or observe them. If you have a large herd, plan to spend several hours on weekends.

For specialized activities such as breeding, showing, shearing, or slaughter you have to spend many hours with your goats. You can plan to do these things on a weekend or during vacation.

TIP

If you have to be gone during the day and have concerns about how your goats will do, find a friend or neighbor who can check in on them or if you can, hire a reliable helper. Develop a backup plan before you need the help, just in case you have an emergency.

Finding and training a reliable helper

Finding a reliable helper can be a challenge. You may find yourself tied to your goats and farm with no time off, unless you have a partner or housemate to help out.

Consider having different people to help you do different parts of your goat care. For instance, you can hire a person with basic skills to do daily chores, but you may want an experienced person to do your milking.

To find and train a good helper:

- Ask neighbors, other goat people, and people at your local feed store whether they can care for your goats or know someone who would be willing to help.

  Some teenagers (and even younger children) are more than happy to help. One time when I had to be gone for a week, my 9-year-old neighbor milked for me. Of course, I had to teach her first. If you have dairy goats, you may be able to find someone who would love to learn how to milk and is willing to care for your goats in exchange for milk.

WARNING

- Make clear to your helper the number-one rule of goatkeeping: Look closely at each goat once or twice a day to make sure he's healthy. I read about a wether that died because the owner's husband had been feeding the goats and didn't notice that the wether was off by himself not eating until it was too late to save him. You don't want this to happen to your goats.
- Do chores with your helper so she can see exactly how much to feed the goats and the steps you take. Point out different goats' behavior, noting individual personality differences among them, so the helper has an idea of what to look for.
- Have the helper do the chores by herself while you watch. Remind her of anything she misses, whether she is giving the correct amount of food, or any other important steps.
- Have the helper fill in or work with you for a day from time to time so she can keep up her skills.
- Be grateful and express your gratitude for the help!

Maintaining Physical Fitness

Maybe you don't have the time or determination to teach your goats to pack or to pull a cart, but they still need to stay physically fit. Goats are into exercise by nature because they're browsers, which means they're always on the move. As long as they have each other and ample space to move around, they can stay physically fit. But why not exercise them in a way that benefits you both?

In this section I talk about some things you can do with your goats to make your life and theirs more fun.

Walking with your goats

Exercising with your goats can be as simple as taking a walk around the pasture or any other convenient place.

I love the joyous, sidewise jumping that kids do when they go down a hill, so you'll sometimes see me running down the hill coaxing the kids to run with me. Sometimes even the adults get into it, bouncing down the hill.

Practice walking with a lead around the pasture and kill two birds with one stone — exercise and training! Or if you're in the city, take them through the neighborhood on a leash (giving a wide berth to the neighbor's prize roses).

Furnishing your yard or pasture with toys

Kids love to play and, like all baby animals, are bursting with energy. You can give them a place to expend that energy by putting goat toys in your pasture.

Some suggestions for pasture toys include

- **Wooden spools, like those used for electrical wire:** Kids love to jump on them. Make sure to cover the hole in the middle or someone will get hurt.
- **Old tires, the bigger the better:** My dad found a huge used tire and immediately thought of the goats. I have videos of them playing king of the mountain, jumping on and off, and running in circles around that tire.
- **Wooden ramp structures:** These are easy to make with a wooden base and a 2 x 6 or 2 x 8 ramp to walk up to the top.
- **Old Little Tikes plastic playground equipment:** You can sometimes find these cheap at garage sales. I have a couple of the picnic tables that my goats love to sleep on and under and a play structure with a slide. I put them indoors in the winter, and the baby goats love to hide and sleep under them.
- **Rocks:** Goats have fun jumping and climbing on large rocks, which offer a secondary benefit of helping wear down hooves.
- **Plastic storage tubs:** Turn the tubs upside-down and kids can jump on them. Put them on their side or right-side-up and kids will sleep in them.

Entertaining with Your Goats

Goats, with their sweet temperaments, make good therapy animals. People love goat kids, but many of them never have a chance to see one. You can help educate people about goats and give them a fun experience by visiting a school, nursing home, group home for people with disabilities, or other institution with your goats.

Pets, including goats, can help people in a variety of ways. Bringing a cute goat to a nursing home can help break the isolation or boredom that some people feel. Goats meet the human need for touch when people get to pet them. They can even bring up good memories from an older person's past, when they lived on a farm or had goats. If you've taught them some tricks, all the better!

I've taken kids to an afterschool program for first graders, and I've also had people with developmental disabilities visit my farm and hold and pet goats, both with positive results and a lot of laughter.

I recently sold a doeling to an enterprising woman who is clicker-training the doeling and another kid to do tricks. She plans to provide their services at birthday parties and other celebrations.

REMEMBER

Unless you've taught your goat to pee on command or use a diaper and you're bringing him indoors, make sure to keep him off carpet, watch closely to prevent accidents, and have a rag or paper towels for quick clean-up. (This is another time you'll be grateful for goats' pelleted poop — easy clean-up!)

You can involve your goats informally or you can go through a formal program. The Delta Society (`www.deltasociety.org`) has a program called Pet Partners that "enables pet owner volunteers to provide services to people in their own communities while spending quality time with their pets."

» Handling hooves

» Disbudding kids

» Castrating bucklings

» Identifying your goats

Chapter 9

Handling Routine Care and Important One-Time Tasks

Goats don't require a whole lot of regular or intense care, but the care they do require is important to their health and happiness. Some of these tasks are a one-time affair, others require regular attention, and many are time sensitive — put off castrating a buckling, for example, and you might end up with kids you weren't expecting.

In this chapter, I show you the proper way to handle and groom your goat, and I walk you through trimming hooves, clipping, disbudding kids, castrating, and tattooing and microchipping.

Grooming Your Goats

Unless you're gussying them up for a show, goats don't require a lot of grooming. (Chapter 17 tells you about showing your goat.) Grooming pays dividends in the long run — by making the goat feel better (who doesn't feel better with a good brushing?), enabling you to evaluate the goat's health, and giving the goat more experience with being handled.

Brushing

Brushing removes dandruff or loose hair that some goats get and increases blood flow — improving the health of the skin and coat. Grooming a goat also gives you an opportunity to check for any signs of illness or disease, such as a lump, swelling, or other abnormality. (For more information on these and other ailments, see Chapter 11.) At a minimum, brush goats in the late spring or early summer, when they're shedding or throwing off the undercoat that kept them warm in the winter. They also are most likely to have acquired lice, ticks, or other little critters that normally take up residence in the winter.

Use a firm-bristled grooming brush like you can get in any feed store or livestock supply catalog. Brush in the direction of the coat starting at the neck, then down the back and down the sides. Make sure to brush the neck, chest, legs, and abdomen.

Bathing

You don't have to bathe goats, but doing so helps remove the lice, make clipping easier, and keep your clipper blades sharp for a longer time.

Goats prefer to be washed with warm water but will survive the inevitable cold water that is all most of us have available. Use a goat or animal shampoo. Caprine Supply has a waterless shampoo that you can use when the weather is cold or when you don't want the bother of bathing.

You can use a collar to secure a baby goat or a goat that you can easily control. Secure other goats on a milk stand or by putting on a collar and attaching it to a fence. After the goat is secure, just wet it, lather up the shampoo, and rinse.

TIP

If you plan to clip the goat right away, blow-dry its hair. Otherwise, let the goat dry naturally. (And hope it doesn't lie down in any poop!)

Clipping

An annual clipping is a good idea for all goats. Shorter hair helps goats stay cooler and allows sunlight to reach their skin, which drives away lice and other critters. Choose a day after the cold weather is expected to be over.

TIP

Use electric or battery clippers from a pet-supply store or a feed store. The Oster A5 or the Andis Stable Pro clipper are good choices. While clipping, check the clippers frequently to ensure that they aren't getting too hot. Spray frequently with a clipper cooling spray or oil as needed. Clean and oil the clippers between each goat, or more frequently, as needed.

If possible, wash the goat before clipping. Clipping a clean coat gives your clipper blades the longest life possible. Watch out for dull blades, which pull on the goat's coat and cause discomfort.

You can clip the beard on a doe for a nicer look, but unless you're going to show a goat, don't bother trimming her head, which is a real challenge. (Chapter 17 tells you how to do it.) Your goat may look a little funny, but a hairy head does her no harm.

Here are the steps to take when you trim your goat:

1. **Secure the goat.**

   Hold a baby goat; put an adult goat onto the milk stand with some grain or secure her to a fence or gate with a collar and a short rope, and give her some grain or hay for distraction.

2. **Start by trimming the top of the body against the grain of the hair.**

   Use a 10 blade on the body. If you want the coat even longer, use a comb attachment. Use long smooth motions to avoid choppy-looking hair. Press on and move the skin over the hip bones and other bony areas for a smooth cut.

3. **Clip the back and each side, and then the legs, neck, and chest.**

   Move from the areas of longer hair to shorter. Use short strokes on the legs and to correct any areas that you missed.

4. **Clip the hair that hangs over the hooves.**

5. **If you're trimming a doe's udder, do so with a 30 or 40 blade.**

   You don't want to take any chances with nicks. Clip to about the middle of the belly and under the legs, lifting one leg at a time to get the sides of the udder. Hold each teat between the thumb and two fingers to avoid nicking as you trim around it. (See Chapter 17 to find out about trimming the udder for show.)

6. **Brush the loose hair off and give the goat a once-over, trimming any uneven areas.**

   You don't have to do a perfect job (and those bald patches *will* grow out).

WARNING

White or light-colored goats can get sunburned. Prevent sunburn by rubbing cornstarch on the exposed areas daily until the hair grows out a bit.

Specialized clipping

Two areas that most people clip more frequently are the tail area prior to kidding and the udder during milking season:

- **Tail to kid:** During and for several weeks after kidding, blood and fluids stick to the goat's tail and the coat around the tail. Clipping the hair a week or so prior to her due date makes cleanup much easier. Clip up the sides of the tail, across the end of the tail to make a short little brush, and around the vulva area and inside top of the back legs.
- **Udder:** Removing hair from the belly and udder makes the udder easier to clean before milking and prevents hairs from falling into the milk. I do this when I clip the doe's tail for kidding and then intermittently during milking season. Clip according to Step 5 in the "Clipping" section.

Caring for Hooves

Keeping a goat's hooves trimmed is one of the easiest, least expensive, and most important parts of goat care. Regular trimming takes very little time and cuts down on health care expenses in the long term. Imagine if you had to walk around with toenails that curled under your toes.

Overgrown hooves can affect the way a goat walks, throwing its body off and leading to leg and foot problems. They can also limit a goat's activity and exercise level, leading in extreme cases to decreased food intake because walking to get to food is so painful. In wet and rainy weather, goats with neglected hooves are more prone to foot rot, which can be a long-lasting condition and create additional problems, as well. (Chapter 11 tells you about foot rot and other common ailments.)

TIP

Always check a back hoof to decide whether it's time to trim. Front hooves wear down more quickly than back hooves because goats use them to paw at things.

How often you trim depends on each goat and the conditions it lives in. I sold a goat to a woman who said that she rarely had to trim her goats' hooves because the goats spend most of their time in gravel. Be creative and include objects that will naturally wear down the hooves — such as large rocks or coarse roofing shingles on top a structure they like to jump on.

REMEMBER

Don't procrastinate! Check hooves frequently and keep your goat used to handling by routinely trimming hooves.

Preparing to trim

Trimming a goat's hooves requires very little equipment — a pair of sharp trimmers will do the job. Livestock supply stores or catalogs offer trimmers (see Figure 9-1) specifically geared to hoof trimming for a reasonable price. Although not essential, a small Dremel sanding tool helps to get the bottom of the hoof nice and flat. I do all the trimming with trimmers specific to the job, or with rose pruners with a straight edge.

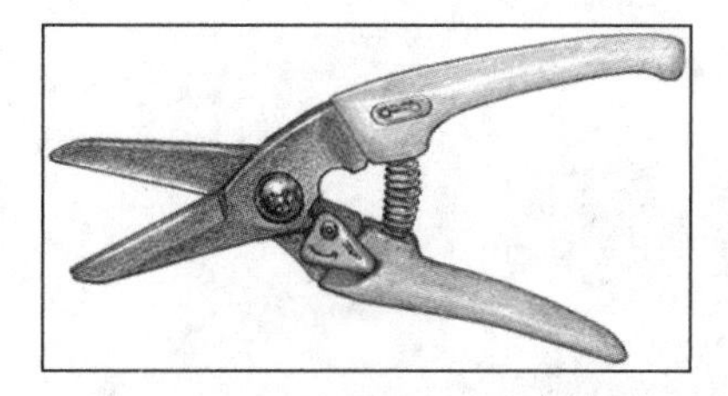

FIGURE 9-1: You can buy hoof trimmers from a livestock supply store.

TIP

For kids, the only supply you need is hoof trimmers. You may accidentally cut too deep, however, and having some blood stop powder may make you feel better. (You can find it at any livestock supply store.) But, as a goat friend of mine once said to me, "No goat has ever bled to death from a cut caused by hoof trimming." Usually, after a goat puts pressure on the foot by standing, the bleeding stops.

You can trim a young kid's hooves simply by holding him on your lap and bending the front foot back or the back leg out. The kid may struggle, but you usually win.

Adult goats, just because of their size, need to be restrained in some manner for hoof trimming. Ideally you use a milk stand, but an alternative is to put a collar on the goat and tie it to a fence or gate with a short rope. Another option, which may be less effective depending on the temperament of the goat, is to have a helper push it up against a wall and hold it in place.

Trimming the hooves

Trim any hoof that is not flush with the base of the foot A hoof that hasn't been trimmed recently folds over the sides. In cases where it has been neglected the hoof will be growing longer in the front and on each side and may be malformed.

WARNING

Do not try to fix a severely overgrown hoof in one trimming. It took time to get that bad and will take time to be corrected.

Follow these steps when the time comes to trim your goat's hooves:

1. **Clean the hoof of manure or other debris.**

   Doing so helps your trimmers stay sharp longer.

2. **Lift the foot to be trimmed, bending it back at the knee, as shown in Figure 9-2 and Figure 9-3.**

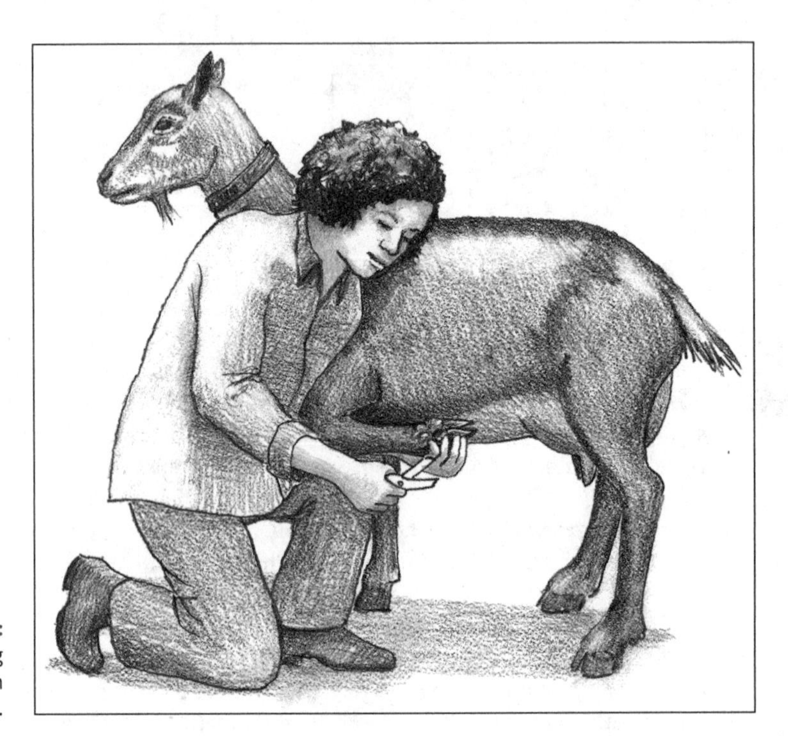

FIGURE 9-2: Bend the leg before you begin clipping a hoof.

3. **Trim the folded over or overgrown sides of the front wall of the hoof one at a time, as shown in Figure 9-4.**

4. **Trim the heel gradually so that it is even with the sole (see Figure 9-5).**

   Be careful not to cut too deep. If you begin to see pink tissue, stop. You can always trim a little more later. Figure 9-6 shows you a finished hoof.

5. **Gently sand the hoof and pad flat with the Dremel tool, if using one.**

6. **Repeat with the other three hooves.**

TIP

After each trim, disinfect your trimmers with alcohol, and use on the next goat or store in a dry place.

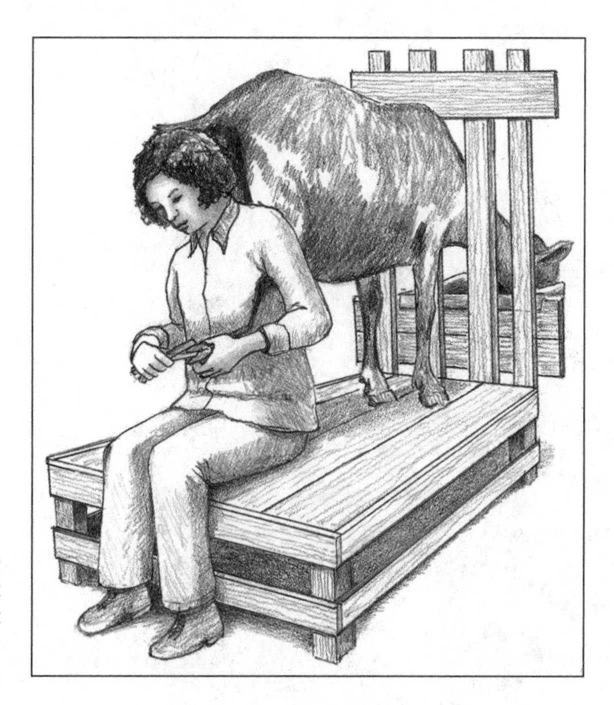

FIGURE 9-3: You can sit behind the goat to trim back hooves.

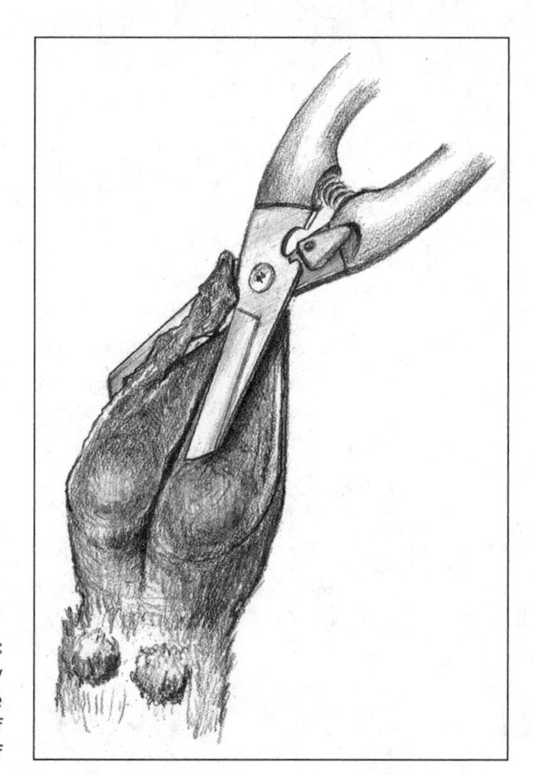

FIGURE 9-4: First trim off any growth on the front walls of the hoof

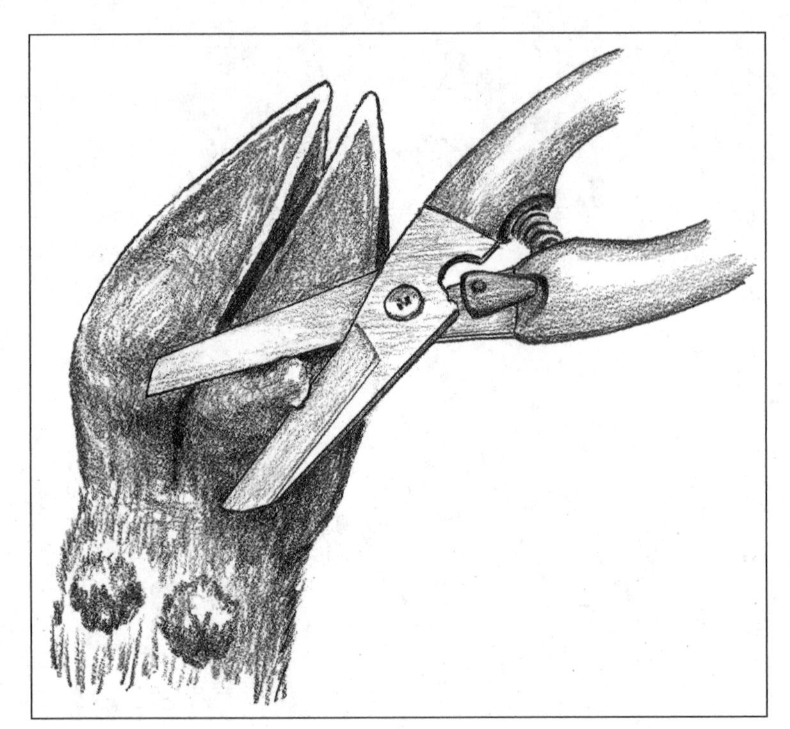

FIGURE 9-5: Trim the heel a little at a time so it is even with the sole.

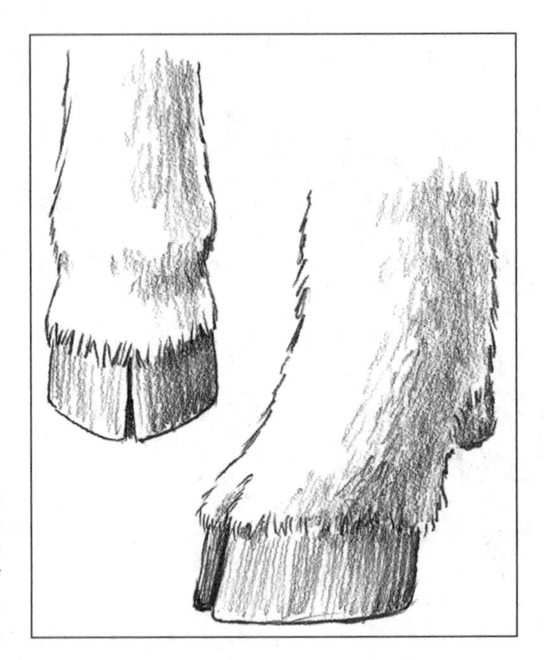

FIGURE 9-6: A goat stands on a properly trimmed hoof at about a 45-degree angle.

Dealing with Horns

Horn care can be a one-time deal — removal of the horn buds, called *disbudding,* shortly after birth. If horns are not removed early, and you want a hornless goat, you need a veterinarian to remove them. The procedure is risky and leaves a temporary hole in a sinus within the goat's skull, which may be slow to heal.

The only other time you need to care for horns is if they break or if further growth, called a *scur,* develops after disbudding — usually in a buck. (See the section "Preventing and dealing with scurs" for more about scurs.)

Horned or not?

Dairy goats are generally disbudded, although a minority of goat owners prefer horns because they believe it is more natural. Fiber goats are often left with horns, which are believed to help with temperature regulation, a consideration for animals with a heavy coat. Some meat goat shows also require that the animals have horns.

Some people like the fact that you can "steer" goats by their horns. To eliminate some of the danger, they grind down the horns to remove the points (called *tipping*). Tipping can cause excessive bleeding, so I don't recommend it.

For those who prefer hornless goats, the ideal may be naturally hornless goats, called *polled* goats. The trait for polledness requires that at least one parent be hornless.

TIP

If you are concerned about or are having problems with a horned goat, try wrapping the horns with foam and duct tape, or putting polyethylene black plastic water pipe over them for padding and protection. (Don't use this method if you see other goats eating the padding, because the materials can cause a blockage in the digestive system.)

TECHNICAL STUFF

In certain goat breeds, the traits of hornlessness and *hermaphroditism* (having male and female sex organs) are linked. Breeding two polled goats to each other may increase the likelihood of producing a hermaphrodite.

The case for no horns on goats

Disbudding is best for the goat and the owner, for a number of reasons:

- Horned goats can get their heads stuck in fences or feeders. I have had to cut out parts of fences when I was unable to free a goat whose horns were stuck. A stuck goat is vulnerable to being butted by other goats while having no way to fight back or get away, or to dying in the hot sun if left too long.

» Horned goats are more destructive. They are prone to rubbing and banging their horns on structures, fences, and feeders. A horned goat can hurt you, your guardian dog, or other goats worse than a disbudded one, particularly if the other goats don't also have horns. (Imagine a horn in your eye, for example. It happens!) You may be liable if your goat injures another person with his horns.

» Because horned goats are not allowed in most goat shows, are more dangerous, and are less popular among buyers, their market potential is less than that of disbudded goats.

» If you want to milk a horned doe, you need a milking stand with a special head lock that horns can fit through.

When to disbud

Kids need to be disbudded within the first two weeks after birth, and for some males, as early as two days old. Females usually do not have as much horn development as males, so earlier disbudding is not as important. Do not put off disbudding for too long, though, because doing so can make the job difficult, cause the goat unnecessary distress and pain, and result in horn regrowth or scurs.

How to disbud

You have to decide whether to use a veterinarian, have an experienced goat friend disbud, or do it yourself.

WARNING

If you decide to have a veterinarian disbud your goats, make sure that he or she has experience with disbudding goats. (See Chapter 10 to get the scoop on finding a goat vet.)

I had a goat friend disbud my first two kids while I observed. For the next round of kids (the next year) I bought a Rhinehart 50 electric disbudding iron with a goat tip and did it myself. You need to make sure that you use an iron with a goat tip, and not a heavy-duty cattle iron, to avoid injuring the kid.

I don't use anesthesia (something that is required in the United Kingdom but not the United States) because even though the kid may feel more pain (less than 20 seconds) without it, the recovery period is only a few minutes. They still vocalize as though in pain with the anesthesia, but then they sleep for at least a half hour afterward. After having tried it on several goats, my veterinarian and I are not convinced that they feel no pain. We agreed that the stress (to us) of worrying about them waking up did not outweigh the benefit (to them) of having no memory of the procedure.

WARNING

Never use a disbudding paste on a goat. Because of their nature, goats will rub the caustic substance on each other, which can lead to chemical burns or even blindness.

Gathering supplies

Unless you disbud kids under anesthesia, a kid holding box is essential. This rectangular hinged box exposes just the goat's head, enabling you to disbud and tattoo without having to hold a struggling body.

TIP

You can buy a kid holding box from a goat supply catalog such as Caprine Supply. You can also buy a metal headpiece and make your own box, or you can make a box with a wooden headpiece (you can find lots of plans for boxes online). Be aware that miniature goats need smaller kid boxes than standard goats.

You also need the following supplies:

- A syringe with 1 cc of tetanus antitoxin which you can get at a feed store. This will protect the kid from tetanus for 10-14 days.
- Disbudding iron from a goat supply catalog (see Figure 9-7).
- Pain reliever. I use the prescription painkiller and anti-inflammatory Banamine, which I get from my vet. Aspirin or ibuprofen also do the trick, but you need to give them with food.
- An antiseptic spray such as Blu-Kote, which you can get at a feed store.
- If the kid is bottle fed, a bottle to comfort it following the procedure.

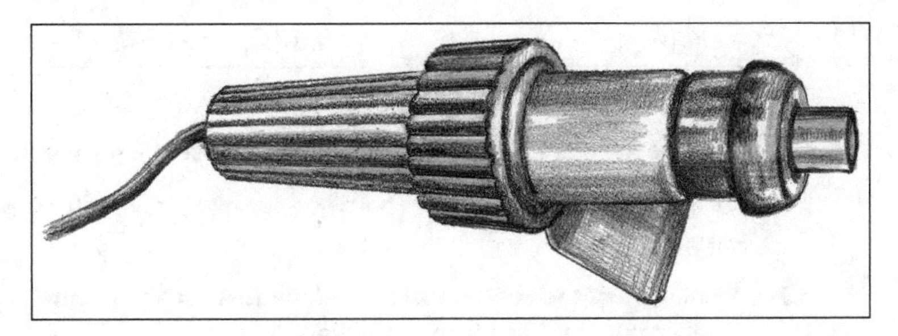

FIGURE 9-7: A disbudding iron burns the horn bud, causing it to eventually fall off.

Disbudding the kid

Follow these steps to disbud a kid:

1. **Preheat the disbudding iron.**

   Heat it until the end is red hot — about 20 minutes.

2. **While the iron is heating, give the kid the tetanus shot and the pain reliever, which takes about a half hour to work.**

3. **Restrain the kid.**

   Position the kid in a kid holding box (see Figure 9-8) so that the ear nearest the horn bud you start with is tucked back into the kid box. You also can recruit someone to hold the kid while you disbud. If you go that route, make sure that person is wearing heavy gloves and a long-sleeved shirt so that she doesn't get burned when the kid struggles.

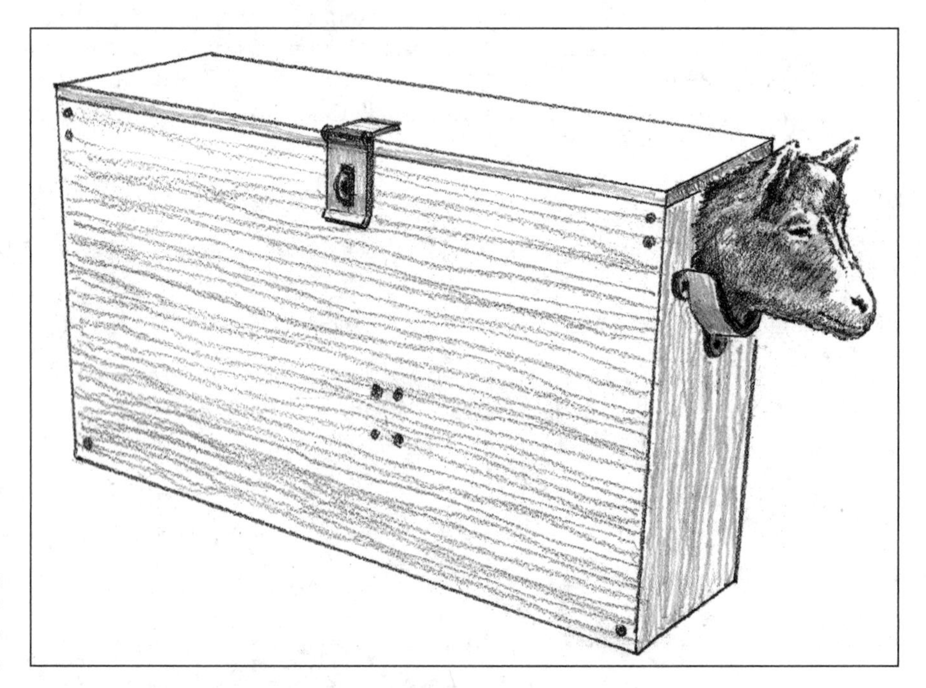

FIGURE 9-8: Kid ready for disbudding in a kid holding box.

4. **Clip the horn area with your clippers to expose the horn bud.**

   Clipping the hair keeps it from burning and smoke going into your eyes as you disbud.

5. **Firmly grasp the goat's muzzle, making sure it can breathe, and evenly apply the disbudding iron to the horn bud.**

   Hold the iron on the bud while applying firm pressure and gently rocking the iron for 8 seconds, keeping the kid's head immobilized (see Figure 9-9). For older kids or bucks who have some horn growth, allow up to 8 more seconds. The kid will struggle and yell, but the process is over very quickly.

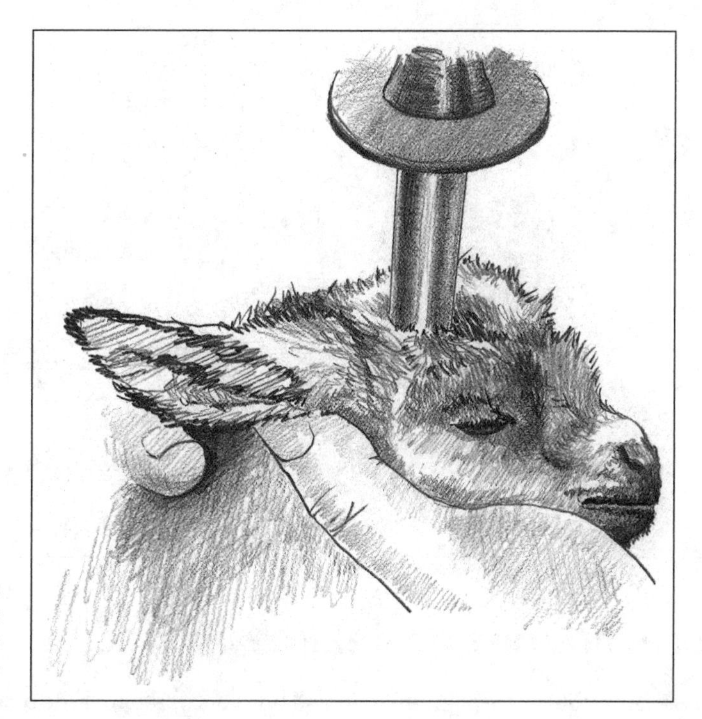

FIGURE 9-9: Hold the kid's head still while you apply the disbudding iron.

6. **Check to make sure that you have left a dry-looking, copper ring around the horn bud.**

   Figure 9-10 shows you how a properly disbudded horn looks. If you don't see a copper ring, apply the iron for only another few seconds. (If you feel you need to burn more than this, give the kid a few seconds break to avoid overheating the skull and possibly injuring the brain.)

7. **Remove the part of the bud inside the copper ring with your fingers. If it bleeds, you can cauterize it by applying the disbudding iron lightly.**
8. **Repeat Steps 3 through 7 with the other horn bud.**
9. **Remove the kid from the box and spray antiseptic spray on the disbudded area, taking care to avoid the eyes.**

TIP

When the process is over, give the kid its bottle or put it under its mother to nurse. Try not to let the doe smell the kid's head or she may try to reject it.

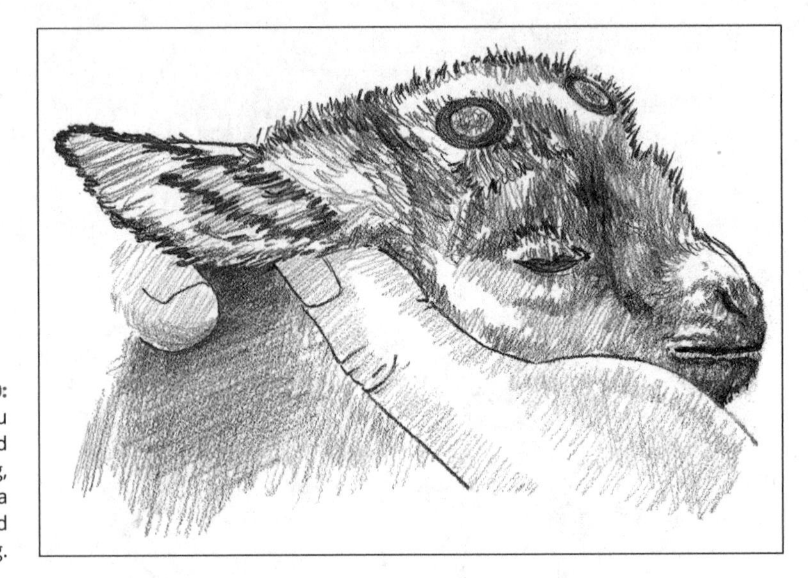

FIGURE 9-10: When you are finished disbudding, you see a copper-colored ring.

Preventing and dealing with scurs

Disbudding burns through the skin and horn bud, stopping the blood supply to the horn buds and causing them to eventually fall off, if all goes well. When they do grow back, the partial horns are called *scurs*. Figure 9-11 shows a goat with scurs.

FIGURE 9-11: Horns that regrow after disbudding are called scurs.

The horns can regrow, particularly in bucks, if they are not disbudded early enough or well enough. Because the horn grows wider at the base as the kid grows, and the growth is faster in bucks, getting all of it can be challenging. Some people burn a circle eight, or two circles, at the time of disbudding to prevent scurs.

TIP

Check for regrowth a few weeks after disbudding and reburn if you see a problem developing.

In many cases, scurs start growing long after the kid is too large to fit in a kid box or handle easily. In those cases, I just let the scurs grow. During breeding seasons the bucks often knock these scurs off while fighting.

Unless they grow exceptionally large, scurs aren't usually a big problem. They may cause a lot of bleeding when they break off. In those cases, I just spray the area with Blu-Kote to help prevent infection and keep an eye on it.

Exceptionally large scurs can lead to the same problems I talk about in relation to horns. (See "The case for no horns on goats.") Another problem I have seen is a scur that grows toward the goat's eye. A friend with this problem resolved it by cutting a child's "water noodle" (used for swimming) into a short piece and gluing it around the scur to hold it out of his eye. In some cases you may need to have your veterinarian cut part of it off every so often.

If the damage from a broken scur is extensive, or if the bleeding won't stop, you should cauterize it with a disbudding iron or contact a vet.

Castrating Your Bucks

Responsible goat owners who breed goats eventually have to turn a *buckling* into a *wether* — that is, castrate him. Fortunately, castration is an easy procedure and is surprisingly not that hard on a goat. With a minor painkiller, a castrated goat is usually back to his old self within a few hours after the procedure. The main decisions you have to make about castration are which bucks to castrate and which method to use.

The problem of poor Elmer, or why to castrate

In a story widely circulated among goat owners, a woman follows the sad life of a buck named Elmer that a family chose not to castrate. They love him as a kid, but then he grows up and is not so cute. They sell him to the next unwitting owner, and so on, until a scruffy, stinking Elmer is sold at auction for almost nothing and ends up alone in a field, tied to a stake and not properly cared for.

The story illustrates one reason you should castrate bucks unless you *know* they will be sold for meat: They outgrow their cuteness and become undesirable as pets. Other reasons you should castrate bucks:

- **You need only one buck to breed many does.** Your herd is in its best shape if you allow only the best of the best to become breeders. Far too many people (especially novices) keep bucks that are not from the highest quality parents.
- **Bucks require a separate living space to keep kids safe and to control breeding.**
- **Bucks can be hard to handle.** They're harder to handle and more likely than does or wethers to become aggressive — especially during breeding season.
- **Bucks stink, literally.** They urinate on themselves during breeding season and have scent glands that put out an aroma that many people find unpleasant.

On the other hand, although wethers can get big, they are the sweetest of all goats, they don't stink, they make great pets and pack animals, and they don't go into heat and make a ruckus like does or bucks.

Knowing when to castrate

The ideal time to castrate a goat is when he is 8 to 12 weeks old, if he is to be a pet or a pack goat. If you know that he will be used for meat you can castrate as early as a week old. Castrating too early can predispose the goat to developing urinary stones. (Chapter 11 tells you more about urinary stones.) Castrating too early may prevent the *urethra* (the passage from the bladder to outside the body) from developing to its full size. Castrating too late can lead to inadvertent breeding — bucklings as young as 2 months old have been known to breed does. A larger animal also is harder to restrain, and castrating late can cause more discomfort or medical problems for the goat.

REMEMBER

Unless your buckling has a future as food or as a herd sire, mark your calendar for eight weeks after his birth and be sure to follow through with castration.

Choosing a castration method

Castration is an easy procedure. If you are squeamish about it, have a veterinarian or a goatkeeper friend do it for you. If you want to observe before trying it yourself, volunteer to hold the goat for the procedure. All of the methods require a helper to hold the goat.

About a half hour before castrating the kid, give him one adult aspirin, 0.25 cc of Banamine (prescription only), or some white willow bark tincture to help prevent pain, and 1 cc tetanus antitoxin to prevent tetanus. The kid will experience some discomfort after castration but will soon forget about it.

Elastrator castration (banding)

Banding is the most common method that goat owners use to castrate their goats. It is quick, easy, bloodless, and reliable. *Banding* refers to applying a small, thick rubber band (called an *elastrator band*) to the top of the testicles with a metal tool called an *elastrator*. Figure 9-12 shows you an elastrator.

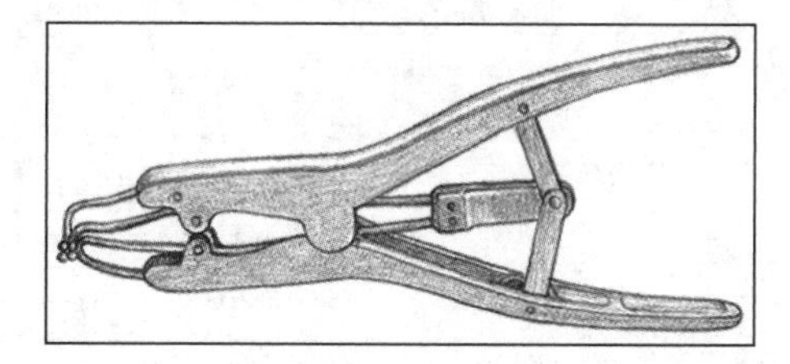

FIGURE 9-12: An elastrator is the most common tool for castrating.

TIP

To prevent bands from breaking down, keep them in the refrigerator until you are ready to use them.

To prepare for banding, place a band that has been soaked in alcohol for a few minutes on the prongs of the elastrator. No other disinfectant or cleaning is needed because the procedure is bloodless.

Follow these steps to castrate using an elastrator:

1. **Restrain the kid.**

   Your helper can hold the kid in her lap, facing outward with his back to her chest, or she can straddle the goat and lift the kid's back legs up so he is standing on his two front feet.

2. **With the prongs of the elastrator facing the kid, expand the band by squeezing the elastrator.**

3. **Place the band over the scrotum and testes close to the body, making sure that both testes are under the ring.**
4. **Release the elastrator and pull it from the band, making sure that the band is close to the body and that the teats are not trapped in the band.**

   If you believe that the band is not on correctly or if one of the testes is not below the ring, cut off the band and repeat the procedure.

The scrotum and testes dry up and drop off in about two or three weeks. Check them regularly after that if they have not fallen off. Check them for infection and spray with Blu-Kote or another spray antiseptic, if needed. In a few cases, they may be hanging by a small amount of tissue, and you can cut them off with a clean scalpel or sharp knife.

Emasculator castration (Burdizzo)

Emasculator castration is also considered bloodless and is done with a somewhat heavy tool called a Burdizzo, or an emasculatome (see Figure 9-13), which crushes the spermatic cord. It is not as reliable as other methods because you cannot see whether the cord has been effectively crushed.

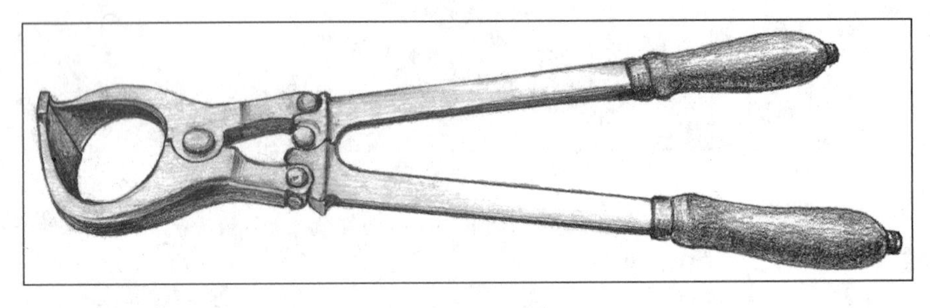

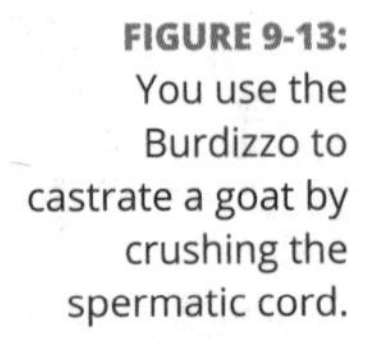

FIGURE 9-13: You use the Burdizzo to castrate a goat by crushing the spermatic cord.

TIP

Never try to crush both cords at the same time. Follow these steps to castrate with a Burdizzo:

1. **Have the helper hold the goat on his lap, with a front and back leg in each hand.**
2. **Feel the scrotum for the cord and move the cord to the outer side of the scrotum.**
3. **Clamp the Burdizzo over the cord, making sure to avoid the teats.**
4. **Squeeze the Burdizzo to a count of 25, and then release.**

   Crushing the cords makes a crunching sound. Squeeze the Burdizzo for a second, shorter time.

5. **Pull slightly on the testicle to ensure that the cord has ruptured.**
6. **Repeat the process on the other spermatic cord.**

The testicles gradually shrink into a small scrotal sac. This method generally leaves no wound but can cause small abrasions.

Surgical castration

Surgical castration is the most reliable and least expensive method, but it is not for the squeamish, leaves an open wound, and is beyond the scope of this book.

Identifying Your Goats: Microchipping and Tattooing

If you live in a state that doesn't require identification, you don't have to permanently identify unregistered goats. If you get a registered goat, it should already have a microchip or tattoo, and if you want to register a goat that is eligible for one of the registries, you will be required to permanently identify it to prove that the goat is who you say it is. So, indirectly, identifying the goat can add to its value.

You also may want to permanently identify your goats even if you aren't required to. You never know when you might have to prove that they're yours — if they get lost or stolen, for example.

Choosing a method

I have tattooed and microchipped goats over the years, and find advantages and disadvantages to both. (Table 9-1 shows you a comparison of microchipping with tattooing.)

REMEMBER

If you plan to register goats, you need to check with the registry you will use to find out what method they allow and what tattoo sequence (letters and numbers) you need to use. (See Chapter 3 for more information on goat registries.)

Basic equipment for tattooing costs less than $100 and only a few dollars a year after that. On the other hand, purchasing microchips and a reader has an initial cost of more than $400.

TABLE 9-1 Comparing Identification Methods

Microchipping	Tattooing
Minor pain	More painful
Clean	Messy — requires ink
Expensive, with ongoing costs	Relatively inexpensive with minimal ongoing costs
Error-free	Easy to make error — tattoo backward or poorly applied
Can cause tumors or migrate, in rare cases	Low risk of medical problems
You need a reader to see the identification number	Identification number is visible but may fade

Tattooing your goat

Supplies for tattooing include gloves to protect your hands from ink, a pair of tattoo tongs, special tattoo ink, and letters and numbers to use in the tongs. You can use black ink on light-colored goats, but green shows up better on dark-skinned goats.

1. **Secure the sequence of letters and numbers you will use in the tattoo tongs.**

   Squeeze them on a piece of paper to ensure that they are in the right order.

2. **Put on your gloves.**
3. **Clean the inside of the goat's ear or tail web (the loose, hairless area under the tail on either side of the anus) with alcohol, making sure that you have the correct ear for the tattoo you are using.**

   LaMancha goats' ears are too small for tattooing, so you need to use the tail web.

4. **When the ear is dry, rub tattoo ink on the inside of the ear or on the tail web.**
5. **Hold the ear out and position the tattoo tongs over the inside, being careful to avoid the veins to minimize bleeding.**

   To tattoo the tail web, position the tongs with the prongs facing the hairless side of the tail.

6. **Puncture the ear or tail web firmly with the tattoo tongs one time, then release.**

   The goat will try to pull away, so make sure to hold securely.

7. **Apply more tattoo ink and rub it in with your finger or a toothbrush to ensure that it fills the puncture.**
8. **Change the tattoo digits to the sequence you will use for the second ear and repeat Steps 1 through 7.**

Microchipping your goat

Microchips come in sterile, individual injectors that look like a large syringe and needle. Each is sealed, has a unique number, and includes several stick-on labels imprinted with the number. The microchips can be read only with a special microchip reader. I use the AVID microchips from EZ-ID in Greeley, Colorado.

The best place to insert the microchip is in the tail web (the loose, hairless area under the tail on either side of the anus). Always use the left side to make finding the microchip easier.

You need a cotton ball, some rubbing alcohol, a microchip in its injector, a microchip reader, and registration papers and/or another form to record the number. (A reader is not required for microchipping, but I recommend having one to avoid the small chance of error in recording the number.)

Here are the steps you take to microchip your goat:

1. **Get your supplies together.**

   Remove the microchip injector from its container, being careful to keep the needle up so the chip doesn't fall out, and scan it. Confirm that the number scanned is identical to the number on the stick-on labels.

2. **Secure the goat on a milk stand or have a helper hold the goat on her lap.**

   If you're using a helper, have her hold the goat with the head to one side, the legs secured between her legs, and her arm wrapped around the goat's side holding the tail up. She can hold the legs with the other hand for more stability.

3. **Clean the insertion area with alcohol.**

   TIP

   If you have a goat that *may* have been microchipped previously, scan the area several times to verify that no chip is implanted.

4. **Insert the needle just under the loose skin for several inches, pressing upward at a nearly parallel angle.**

   Press the plunger until it stops.

5. **Remove the needle and apply pressure for a few minutes at the injection site to prevent the microchip from coming out and to stop any bleeding.**

6. **Scan to locate the implanted microchip.**

   Verify the number against the stick-on labels. Place a label on your form and registration papers, if applicable, and record the animal's name.

3 Managing Goat Health and Breeding

IN THIS PART . . .

Find a veterinarian, put together a first-aid kit, perform basic health-care procedures, and more.

Keep your goats free of parasites, recognize potential health problems, and prevent nutritional deficiencies.

Breed your goats and know what to expect from goat pregnancy.

Raise kids like a pro.

Know what to expect as your goats age and recognize when it's time to let your goat go.

» **Putting together a first-aid kit**

» **Performing basic health-care procedures**

» **Using records to identify and track health problems**

Chapter 10
Outlining Basic Health-Care Requirements

Raising goats requires more than just putting them out in a field to eat brush and fend for themselves, or putting them in a barn with an automatic waterer and throwing in a bale of hay every so often. In addition to routine care such as hoof trimming and clipping, you need to be ready for any health-care issues that might crop up.

In this chapter I run through points you need to know to take care of your goats' health and be ready for the illnesses, accidents, and surprises your goats are sure to throw at you. I talk about how to find and work with a veterinarian and what to include in a first-aid kit. I show you how to do simple procedures and how to keep health-care records. I also run through some routine tests you might want your goats to have and tell you about legal issues related to giving medications.

Recognizing Signs of Illness

Goats are creatures of habit. You can learn these habits and use them to identify illness simply by observing your goats a couple of times each day. Besides, it's a good excuse to spend time with your goats. (See Chapter 2 for signs of a healthy goat.)

Some goats always stay with the herd, while others tend to go it alone or hang out with just one buddy. When a social goat isolates itself or a loner goat suddenly gets into the middle of the herd and starts fighting a lot, you have a clue that something might be wrong.

A change in eating habits gives you another clear sign. Goats exhibit only minor variations in eating — some are always pigs while others eat more slowly or have to fight or be sneaky to get their share. When a goat stops eating and drinking, you know it isn't feeling well. On the other hand, when a goat starts eating a lot, it's pretty obvious that the goat is feeling great!

Here are some other signs that a goat might be sick:

- **Not ruminating:** Cud-chewing (called *rumination,* see Chapter 2) is a part of how goats digest their food. Healthy goats ruminate after they eat. When a goat stops ruminating, it's a sign that the digestive system is upset.
- **Walking difficulty:** A limp indicates a possible injury or a hoof or knee problem, while staggering alerts you to a possible neurological problem.
- **Teeth-grinding or head-pressing:** Both of these are signs that the goat is in pain and you need to investigate further.
- **Changes in breathing:** Some health problems can cause fast or labored breathing, while others cause the goat to breathe more slowly. Extreme heat can also cause labored breathing in a healthy goat.
- **Cough, runny nose, or runny eyes:** A healthy goat usually has no cough, a moist nose, and dry eyes.
- **Abnormal poop:** Goats normally have firm, brownish, pelleted poop. Changes in consistency or color may signal a health problem.

Whenever you find a clue that something might be wrong with a goat, you need to examine that goat to see whether it has any other symptoms, take its temperature (see Chapter 2), and try to determine whether a problem is developing.

Working with a Veterinarian

Unless you are a veterinarian yourself, you need to plan on working with a veterinarian to handle goat health-care issues that come up. Even if you're competent in some areas because of experience with dogs or other livestock, the time will come when you need to call a veterinarian for assistance.

Finding a vet

Finding a vet who works with goats and has expertise in goat health may be difficult, but you need to find one long before you have a problem. A good place to start is with veterinarians who work with horses or other livestock. Look in the yellow pages or ask other goat owners to find such vets in your area.

If you live in an area with a veterinary school, that's another good option. Not only will you get veterinary care for your goats, but you'll be teaching students and possibly influencing another person to be a goat vet. Another benefit of living near a veterinary school is that you may be able to obtain free or discounted services because of the learning opportunity they provide for students.

Where I live, one of the local equine vets has been willing to work with goats and is fast becoming a goat expert because the word has spread and many goat owners in the area have begun to call him.

WARNING

Don't wait until your goats are sick before you try to find a veterinarian for them. You may not be able to find a vet who will take care of them in an emergency situation and experience a dire outcome.

In order to be most effective, you need someone that you're comfortable with and with whom you can work as a team to provide the best care for your animals.

Finding a veterinarian for your goats is usually not as easy as finding one for a cat or a dog, particularly if you live in an urban area. Treating goats just isn't as lucrative as treating dogs, cats, or horses. As a result, most veterinarians you find know very little about goats. But think of it as an opportunity to teach.

I have had some luck working with a naturopathic, home-care vet, too, so if no livestock vets are available in your area, talk to one of these alternative practitioners to see whether they would be willing to work with your goats and to learn more about them. Doing so benefits not just you and your animals but other goat owners in the area.

Knowing when to call

A good time to make that first call to a vet and to start developing a relationship is when you want to have blood drawn for your herd testing. This first step makes your goats her patients and is likely to make her more responsive when your goats have a problem. Continue to use your veterinarian for routine care that you can't perform, or ask her to teach you to do certain procedures when she is out on a farm call.

When you see a medical problem in your goats that you can't resolve or you believe is serious, don't hesitate to call your veterinarian. The veterinarian may be at another farm, and you may have to wait to get assistance for your goat. If the problem turns out not to be so serious, you may be able to get advice or a prescription over the phone. The next section tells you about the kinds of information you need to communicate a problem to the vet.

Preparing for a vet visit

Before you call a vet to come to your farm or bring a goat in for a non-routine care visit — unless it is a serious emergency — take a few steps to make sure that your goat gets the most appropriate care.

You've already noticed that the goat is not acting like itself. Now you need to write down what you have observed. Make notes of the goat's symptoms, how long it has been sick, and the medications or other care you've given so far. Sometimes remembering everything is hard when you're under stress, and having this kind of information to share helps the vet make a correct diagnosis. (See "Keeping Health Records" for tips about recordkeeping.)

If you have time, do the following before your vet visit:

- Take the goat's temperature.
- Check its gums or eyelids for color.
- Listen for heart rate and ruminations.
- Note whether the goat has
  - Injuries.
  - Crusty eyes.
  - Breathing problems or coughing.
  - Diarrhea.

» Check for dehydration by pinching the skin on the neck in front of the shoulder, using your thumb and forefinger. Note whether the skin snaps back to its normal position quickly or stays in a tent before it slowly goes back to normal. A slow return to normal indicates that the goat is dehydrated.

Record all of your observations for the vet's reference. Also be ready to share the goat's history of prior illness, vaccinations, and other health-care information.

TIP

If the vet will be making a farm call, ask whether you can do anything before he arrives. For example, he might want a urine or fecal sample. You also need to catch the goat and put him in a confined, lighted area while waiting for the vet to arrive.

FINDING HEALTH-CARE SUPPORT ON SOCIAL MEDIA

When I first got into goats, I learned about several email groups made up of long-time goat owners who provided support for and answered questions about goat health issues. With the gradual transition to reliance on social media, such as Facebook, these email groups have gradually died off.

The new Facebook groups can be helpful when you have nowhere else to go with your questions, or just want to see how other goat owners handle certain issues. But plenty of misinformation abounds in these groups — because everyone has an opinion.

When looking for a group to join — or more important, to take advice from — consider who they're holding out as an expert. Ideally the experts giving advice will be veterinarians or long-time goat owners, who in some cases may know more than vets about some issues. Also remember that they should never take the place of veterinarian in an emergency.

Among my favorites Facebook goat health groups are

- **Advanced Goat Health and Care (**`www.facebook.com/groups/AdvancedGoatHealthAndCare`**):** This group is made up of longtime goat owners who have dealt with many issues.
- **Goat Veterinary Consultancies (**`www.facebook.com/goatvetoz`**):** Dr. Sandra Baxendell, an Australian veterinarian, can help with problems and questions.
- **Goat Vet Corner (**`www.facebook.com/groups/goatvetcorner`**):** Dedicated to vets who answer questions about goats. Only vets may answer.
- **Goat Fertility, Reproduction, and Neonatal Issues (**`www.facebook.com/groups/505611549610613`**):** A breeder-to-breeder information exchange.

Working with a non-goat veterinarian

Most goat owners find themselves using a veterinarian who has little or no experience with goats. Most of these vets are willing to admit to this lack of knowledge and work with the owner to learn together.

You can assist in this learning process by

- » Sharing goat health-care information that you get from goat conferences, magazine subscriptions, and other breeders.
- » Suggesting that the vet buy one of the goat health books on the market, or if you can afford it, buying one as a gift.
- » Giving the vet a membership to the American Association of Small Ruminant Practitioners (www.aasrp.org), which has a regular newsletter that always contains current health information.
- » Getting a group of goat owners together to split the cost of a veterinarian's visit. Some people do this with blood-drawing for testing, for example.

REMEMBER

Don't be afraid to share your observations and suggestions with your vet, regardless of her area of specialization. Goat health care is changing all the time as new information becomes available. The only way we can all learn is to share.

Building a First-Aid Kit

If you have only a couple of goats, you probably can afford the occasional veterinary visit. But as your herd grows, you're likely to find that you want to save money and hassle by treating some of their minor ailments or handling some of the health care yourself. But even if you don't want to take over some of this care, you still need to be prepared for those times when a vet isn't available or the problem is minor.

The following lists show you what to have on hand in a goat first-aid kit. You can get all of them from a feed store, a drugstore, or a livestock supply catalog. None require a prescription.

Include the following equipment and supplies:

- » Surgical gloves
- » Drenching syringe for administering medications and electrolytes

- Cotton balls
- Gauze bandage
- Alcohol prep wipes
- Elastic bandage
- Digital thermometer
- Syringes and needles
  - Tuberculin needles and syringes for kid injections
  - ½-inch 20-gauge needles and syringes of various sizes — 3 cc, 6 cc, 15 cc
- Tube-feeding kit (tube and syringe) for feeding weak or sick kids
- Small clippers for shaving around wounds
- Sharp scalpel
- Sharp surgical scissors

Include these medications:

- 7 percent iodine
- Terramycin eye ointment for pinkeye or eye injuries
- Antiseptic spray such as Blu-Kote for minor wounds
- Blood stop powder, for hoof trimming injuries
- Di-Methox powder or liquid for coccidiosis or scours
- Epinephrine, for reactions to injections
- Kaolin pectin or Bio-Sponge, for scours
- Antibiotic ointment, for minor wounds
- Aspirin, for pain
- Activated charcoal product, such as Toxiban, for poisoning
- Children's Benadryl syrup, for congestion or breathing problems
- Procaine penicillin, for pneumonia and other infections
- Oxytetracycline, for pneumonia, pinkeye, or infections
- Tetanus antitoxin, to prevent tetanus when castrating or for deep wounds
- CDT antitoxin, for treatment of enterotoxemia (see Chapter 11)
- Anti-bloat medication for bloat

You also want to include these items:

- Betadine surgical scrub, for cleansing wounds
- Probiotics, such as Probios or yogurt with active cultures
- Powdered electrolytes, for dehydration
- Fortified vitamin B, for goat polio or when goat is off feed
- Hydrogen Peroxide, for cleaning wounds
- Rubbing alcohol, for sterilizing equipment

The Straight Poop: Fecal Analysis

The best way to find out whether your goats have parasites is through *fecal analysis*, which involves collecting a sample of goat berries, mixing them with a solution and viewing the resulting sample under a microscope to see whether they contain too many parasite eggs.

Unfortunately, most goat owners rarely have their goat feces analyzed. For years, the common practice was to regularly deworm goats, sometimes as often as every month, and to rotate the dewormers used to treat the parasites. This led to trouble — the parasites became resistant to available dewormers in some areas of the United States. Rather than just blindly treat for parasites, the better solution is to analyze the feces to determine whether a problem really exists.

TIP

Unless you have a problem with parasites, twice-a-year testing is adequate. One of the most important times to test your does is right after they kid. The stress of the birth can make them more susceptible to parasite problems. If you cannot test your does at this time, deworm them as a preventive measure. (See Chapter 11 to find out how to deworm a goat.)

I also recommend randomly testing some of the kids during their first six months. They're more likely to have high numbers of parasites during this period because their immune systems are not yet fully developed.

All you have to do for fecal testing is collect a few goat berries, put them in plastic bags, and take or ship them to a veterinarian or veterinary lab for analysis. Ask your vet what he prefers and what paperwork is required. Some veterinary labs may require a referring veterinarian, but others allow you to print out the paperwork from their website and submit your own samples.

In my early years of raising goats, I sent samples of feces from each small group of goats that lived in different areas of my farm — for example, the bucks, the does, and the kids. This group testing method is not as precise as individual testing but can give you an idea whether trouble is brewing, and it's much less expensive if you have a large herd. Eventually I bought a microscope and began doing my own fecal testing. I now test when I see a problem with anemia, diarrhea, or even a rough coat, or randomly on different goats in different areas of the farm.

TIP

To do your own fecal tests, you need a microscope that has at least 40X power (you can get one for less than $100 and a few other supplies from American Science & Surplus (`www.sciplus.com`). Fias Co Farm (`www.fiascofarm.com`) has detailed, step-by-step instructions for testing, and it has photos of the different parasite eggs.

After you have determined what kind(s) of parasites are afflicting your goats, you need to deworm them. Only deworm goats with a parasite overload. Table 10-1 lists the dewormers that are effective against specific parasites. Whenever you deworm, always give the dewormer orally at twice the cattle dose (by weight). Don't rotate dewormers. Experts now advise using two dewormers, from different chemical families, one after the other. Combinations of these are available in Australia but not the United States. Chapter 11 gives you specifics on deworming a goat.

TABLE 10-1

Dewormers for Specific Parasites

Parasite	Dewormers
Roundworm, such as barber pole worm, brown stomach worm	Ivermectin, Morantel, Safeguard, Valbazen*
Liver Fluke	Ivomec Plus, Valbazen*
Lungworm	Ivermectin, Safeguard
Meningeal worm	Ivermectin, Safeguard
Tapeworm	Safeguard, Valbazen*

****Warning:*** *Giving Valbazen (albendazole) during the first 45 days of pregnancy can cause abortion.*

Another kind of testing related to parasites is called the Drenchrite test, which can determine which dewormers will be effective. This test is conducted only at the University of Georgia Parasitology lab, and it's expensive. This test came about in response to the serious problem with parasites and resistance to dewormers that has evolved in the southern United States. If you have a problem getting control of parasites in your goats or dewormers aren't working, you can get information about this test at the American Consortium for Small Ruminant Parasite Control (`www.wormx.info`).

Giving Injections

You can have a vet visit or take your goats to a clinic to receive vaccinations or other injections, and many goat owners do this. But others prefer to save the money and do it themselves. Being a hands-on type of person, I learned to give my goats injections in the first year of owning them.

Giving injections is easy after you get over any fear you might have. I recommend you have an experienced person, such as a vet or another goat owner, demonstrate the technique before you try it. You can also practice by injecting into an orange — just remember to dispose of your practice needles and syringes.

WARNING

Anaphylaxis is a severe, sudden allergic reaction. The faster it occurs, the more severe it is. If a goat unexpectedly collapses or goes into shock after an injection, administer epinephrine immediately. The dose is 0.5-1.0 cc per 100 pounds.

TIP

You can get needles and syringes at a feed store, veterinary office, or livestock supply catalog. I usually use tuberculin syringes for kids and 20-gauge needles for adults. Ask your vet whether you need a different size needle.

The two most common types of injections are subcutaneous (SQ), which is just under the skin, and intramuscular (IM), which goes into the muscle. Read the instructions that come with the medication you're using to determine what type of injection to give. *Most* injections that can be given IM also can be given SQ; consult with your veterinarian regarding which injections you need to give IM. I prefer to give subcutaneous injections because then I don't have to worry about hitting a blood vessel or vein.

When you're ready to give injections yourself, gather your supplies and bring your goat into an area away from the other goats. Have someone hold the goat and find the site where you will give the injection (see Figure 10-1). The best places to give injections are the sides of the neck and "armpit" area just behind the front leg. Never use the rear leg because you can hit a nerve and make the goat lame.

Gather your supplies:

- Medication
- Disposable needle and syringe
- Alcohol and cotton balls or other wipes
- Container for sharps (used needles), which you can purchase at a drugstore

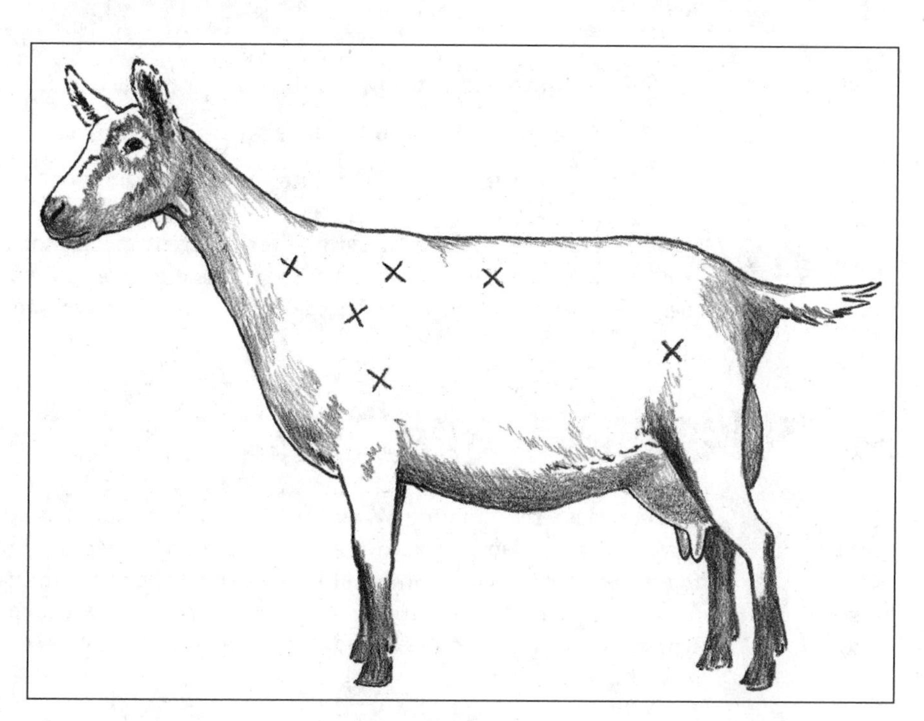

FIGURE 10-1: Injection sites.

Before giving the injection, wipe the top of the medicine vial with alcohol to ensure that it's sterile. Then insert the needle into the bottle and withdraw the required dose of medication. Withdraw the needle and tap the syringe and push the injector slightly to push out any bubbles.

To give an SQ injection:

1. **Lift the skin into a tent.**
2. **Insert the needle under the skin into the tent, toward the body.**

   Make sure that the needle isn't *in* the skin or muscle, or through the other side of the tent.
3. **Inject the medication and remove the needle.**
4. **Discard the needle and syringe into your sharps container.**

To give an IM injection:

1. **Insert the needle into the muscle, being careful not to hit bone.**
2. **Withdraw the plunger slightly to make sure that you have not hit a vein or vessel.**

   If you see blood in the syringe, pull out the needle and start over.

3. **Depress the plunger slowly, and then withdraw the needle.**
4. **Rub the injection area gently but firmly to distribute the medication.**
5. **Discard the needle and syringe into your sharps container.**

Never use the same needle or syringe for different medications or on different goats. Mixing medications can make them less effective or cause an unexpected chemical reaction; sharing needles can spread diseases from one goat to another.

Considering Vaccinations

Vaccines cause the immune system to produce antibodies to a specific disease. You're probably familiar with the flu vaccine and other vaccines that children routinely receive. Some are more effective than others and some, such as the sore-mouth vaccine, actually can give a milder form of the disease. It depends on the type of vaccine and whether it contains a live form of the disease it was created to prevent.

In this section, I give you insight into whether to vaccinate. I tell you about vaccines commonly given to goats and offer pointers for giving a vaccination.

Deciding whether to vaccinate

Most veterinarians recommend that, at a minimum, you vaccinate goats for *clostridium perfringens* types C and D and tetanus (CDT). This vaccine prevents tetanus and enterotoxemia that's caused by two different bacteria. Yet many breeders don't vaccinate their goats with this or any other vaccine, for different reasons.

Vaccinating for enterotoxemia or another disease doesn't always prevent the disease. But in some cases, if a vaccinated goat does get the disease, it will be shorter and less severe, and the goat is less likely to die. And the cost of vaccinating is minor, in comparison to treating the disease or paying to replace a dead goat.

Looking into common vaccinations

A number of vaccines are used to prevent disease in goats (see Table 10-2). Most of them are approved for use in sheep but not goats. That doesn't mean that they aren't effective or can't be used in goats but that they haven't been formally tested on goats.

TABLE 10-2 Goat Vaccinations

Vaccine	Disease Protected Against	When to Give
CDT	Enterotoxemia and Tetanus	Does: Fourth month of pregnancy
		Kids: 1 month old and one month later
		All: Booster annually
Pneumonia	Pasteurella multocida or Mannheimia Haemolytica pneumonia	Two doses 2-4 weeks apart
CLA	Cornybacterium pseudo-tuberculosis	Kids: 6 months old, 3 weeks later and annual booster
Rabies	Rabies	Annually
Chlamydia	Chlamydia abortion	First 28-45 days of pregnancy
Soremouth	Orf	Annually

Most goat owners with small herds usually don't need any vaccines other than CDT and tetanus. In areas where rabies is rampant, some veterinarians recommend that you vaccinate your goats for rabies, even though it isn't approved for goats. I suggest that you work with a veterinarian to determine what is right for your circumstances.

Giving a vaccination

All goat vaccines are formulated to be and so must be given as injections. Follow these guidelines when giving a vaccination:

- To minimize the chance of an adverse reaction, vaccinate goats only when they are in good health.
- Do not use expired or cloudy vaccines.
- Use a 20-gauge, 1-inch or ¾-inch needle on an adult, or a ½-inch needle on a kid.
- Follow the manufacturer's instructions for dosage.
- Use a new, sterile needle and syringe on each goat.
- Do not mix vaccines.
- For the best effect, do not delay booster shots.
- Keep a record of vaccinations given.

Keeping Health Records

Documenting the care and treatments that you or a veterinarian provide is an important part of keeping your goats healthy. The form of the record is less important than the information that you record. You can even write on a piece of notebook paper if that works for you.

With my herd down to fewer than ten goats in the last few years, I've begun to document hoof trimming, deworming, breeding, kidding, and other treatments or procedures on a white board in the milk room. It's big enough to contain everything I've done over the year. At the end of the year, I take a picture of the board, and — *voilà!* — I have my annual record ready to be printed out.

Documenting important information

When I had a large herd, with lots going on, I used two kinds of documents to help me record all the necessary information for routine and critical goat health care: a health-record form and a calendar. I created my own health-record form and I used a wall calendar to supplement it. I recorded current detailed information on the health-record form and current general information (such as the fact that a doe kidded) and future scheduling (for example, the next date for hoof trimming) on the calendar.

Some information to include on a health-care record is:

- Goat's name
- Date of birth or age
- Identification number
- Condition being treated
- Treatment given (hoof trim, injection, and so on)
  - Date of treatment
  - Details or comments
- Kidding information
  - Breeding date or expected due date
  - Sire
  - Kidding date
  - Number and sex of kids
  - Outcome/comments

You can make your own health-care form using this information or other things you consider important. You can also talk to other goat owners to find out what they use or get a form online and use it as-is or modify it to your needs.

TIP

Keep your goats' health record in a place where you can easily refer to it. I kept mine in a notebook in the barn. I still keep the calendar on the wall in the house so I can keep an eye on expected kidding dates and scheduled treatment and care.

Unless you routinely get a goat out for milking or another purpose, you may not think about hoof-trimming until the goat's hooves are overgrown, or giving booster shots within the recommended time frame. The combination of a health-record form and calendar or a white board right in the barn will tell you what you have done and when you next need to do it.

Keeping track of recurrent problems

I had a really sweet goat named Katharine who had a problem kidding her first time — the buck kid got stuck and was dead when I got it out, and the doeling died shortly after birth. The next year she had itty-bitty twin girls that survived. The year after that she cost me some money when another kid got stuck and had to be removed after it died. I sold Katharine that year because when I checked my old records I realized that she had a serious problem. It wasn't just an aberrant condition that first year.

If you're raising goats to make money, knowing which ones are repeatedly getting pneumonia or parasites is critical to determining which to keep and which to sell. If you aren't, the information can still tell you that you might need to change the way you manage that particular animal.

I don't know about you, but I can't always rely on my memory to know which goats get sick every few years. After they improve and do well for a while I forget that they were even sick. Documenting recurrent problems is one of the most valuable reasons to keep accurate health records on your goats.

Providing information for the vet

If you have a goat with an illness that seems minor, you will probably initially treat it by yourself or in consultation with a veterinarian. Using a health-care record to document everything you do makes for a smoother transition if the goat's condition worsens and you have to call a vet for help.

Show your records to the vet so he knows what you have done and what may not be working. These records may even save you some time and money.

Tracking trends

Do you know if a certain doe always has her kids at night? Does your herd always have a higher rate of coccidia during a certain month? Do does in a specific family line always have high numbers of multiple kids? These are just some of the trends that you can discover by keeping health records over a number of years.

Some of these trends might just be interesting, but others help you keep your goats healthier. For instance, you might decide to give a preventive anticoccidial (see Chapter 11) during the month in which more kids got it. Or you may want to evaluate and improve your feeding program for does that usually have high numbers of kids.

Keeping goat health-care records might take a little more time, but they pay off in the long run. They help you schedule care, assist your veterinarian, and provide a wealth of information on individual goats and the whole herd.

Testing to Avoid Problems

Even if you got your goats from a reputable breeder, you want to be sure they don't have a disease unknown to that breeder. This can happen, for example, when the test for a disease shows a false negative or cannot be detected until a kid is older. The best way to make sure they don't actually have a disease is to routinely test them, usually annually.

If you're bringing home your first goats, you need to consider routine testing for common diseases. If you're adding to your herd, the first 30 days when they are in quarantine is the perfect time to have them tested to make sure they aren't carrying a hidden disease. (See Chapter 7 for more about quarantining.)

Unless you have a *closed herd* (your goats never leave and no new goats come in), you need to consider testing or retesting at least annually. Doing so not only gives you peace of mind, but if you sell goats, test results give buyers written proof that the goats are healthy. The upcoming sections tell you about the most common diseases and the kinds of testing you might do to detect them.

Knowing what to test for

You can test for most of the more serious contagious diseases with a simple blood draw. These tests usually are part of annual care. But you can also do them when you have a goat that is showing symptoms of one of these diseases and you and your vet haven't been able to determine the cause.

One test that only has to be performed one time is the G-6-S test for a defect that occurs only in Nubians and Nubian crosses. Testing all kids will help you avoid breeding those with the problem. (I cover this defect in Chapter 3.)

Work with your veterinarian to determine which diseases to test for in your goats. Some tests are expensive and not 100 percent accurate. Chapter 11 tells you more about common diseases.

The diseases that goat owners most frequently test for include

- Three different tests (listed in order of reliability) are used to detect antibody to the **caprine arthritis encephalitis virus (CAEV)** — the AGID, the ELISA, and the PCR. These blood tests don't detect the disease, but a goat that has antibodies to CAEV is considered to be infected.
- The test for **caseous lymphadenitis (CLA)** requires lab analysis of pus that has been aspirated from an active abscess. The CLA vaccine can cause false positive results.
- Several tests for **Johne's disease** are available — some use blood and others use feces. Testing is expensive and not always accurate. Because the disease is believed to be spreading in goats because of goat owners feeding cow colostrum to kids, testing is becoming more common, particularly when the owner has goats that die after a long, wasting disease.
- **Brucellosis** testing is most often done on blood, although a newer test has been developed using goat milk.
- The test for **tuberculosis** is a skin test for antibodies.

How to draw blood for a test

A veterinarian will come out to your farm to do most kinds of tests on your goats. But you can save money by drawing blood from your goats and sending the samples directly to a lab. Ask your veterinarian or another breeder who is comfortable with drawing blood to show you how they do it.

Your veterinarian or another breeder can also help you find out where to send samples from your area and how to ship. To get an idea of shipping requirements, see the Washington Animal Disease Diagnostic Laboratory (WADDL) information at `www.vetmed.wsu.edu/depts_waddl`.

To draw blood, you need a helper to hold the goat and the following supplies:

- Alcohol prep wipes
- 3 ml syringes with ¾-inch 20-gauge needles, one for each goat
- Vacutainer tubes (from a veterinary supply store or vet)

  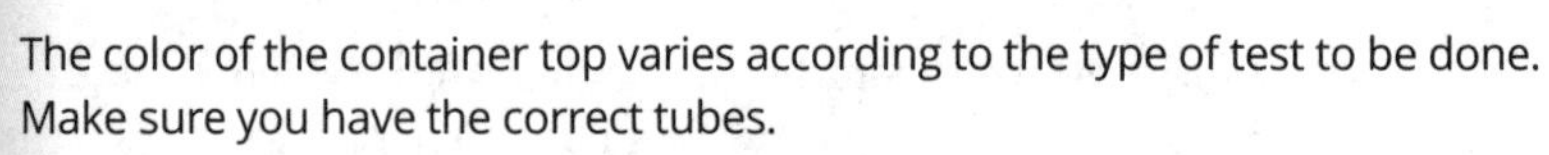

  WARNING

  The color of the container top varies according to the type of test to be done. Make sure you have the correct tubes.
- Clippers, if you need to shave the area
- Paper, pen, and permanent marker

Here's how to draw blood for testing:

1. **Make a list of all goats to be tested, numbering each one.**
2. **Label the tube with the name of the goat, the date, and your name or farm name.**
3. **Have the helper back the goat into a corner and hold the goat's nose with one hand and around the chest with the other.**
4. **Find the jugular vein by pressing on the left side of the goat's throat near the bottom of the neck (see Figure 10-2).**

   The vein pops up slightly when you press on it. If you need to, shave the area to more easily locate the vein.

FIGURE 10-2: Feeling the vein and inserting the needle.

5. **Remove the needle cap and insert the needle upward into the skin and vein at an angle nearly parallel to the vein.**

   Be careful not to the push the needle through the vein.

6. **Gently pull back on the plunger.**

   If blood does not enter the syringe, remove the need and start over.

7. **If you see blood in the syringe, continue pulling the plunger until you have 3 cc.**

8. **Remove the needle, replace the cap, and put pressure on the goat's neck for 30 seconds.**

9. **Remove the needle cap, insert the needle into the goat's labeled tube, and inject the blood.**

When you have finished drawing blood from your goats, refrigerate the samples or prepare them for delivery to the lab or veterinary office.

Knowing the Law Regarding Drugs in Food Animals

Giving medications to goats is not the same as giving them to dogs or cats. By law, goats are considered food animals, even if you keep them as pets. So if you are going to be administering any drugs to your goats, you need to be aware of the law.

The law on prescribing medications for food animals requires that

- You use prescription drugs for your goats only on a veterinarian's order.
- The veterinarian who prescribes the drugs has a veterinarian/client relationship with you prior to prescribing.
- Drugs prescribed for your goats are properly labeled.
- The veterinarians keep appropriate records regarding the drugs prescribed.

Certain drugs are prohibited in food animals, and others that are used in goats are not approved by the FDA for use in goats. (Goats are considered a "minor species," and so very few drugs have been tested on them.) This doesn't mean that a veterinarian can't prescribe them. When the vet prescribes these drugs, it is called *extralabel use,* and the vet has some responsibility for ensuring that they don't leave residues in food.

Each goat owner is responsible for making sure that the meat and milk from her animals don't contain any drug residues. This responsibility, under Food and Drug Administration (FDA) policy, applies to anyone in "the production and marketing chain."

Each medication has a specified *withdrawal time* for meat and for milk. This is the period of time during which the milk or meat from the animal still contains drug residue. You need to be aware of the withdrawal times of all drugs you give your animals, including pain medications and dewormers.

Ask your veterinarian about the withdrawal times for any medications prescribed or that you are unsure of, and make sure not to drink or sell the milk, or sell an animal that might be eaten within the withdrawal time period.

» Learning about parasite management techniques

» Recognizing potential health problems

» Preventing nutritional deficiencies

Chapter 11
Addressing Common Health Problems and Ailments

Even healthy goat herds can develop problems from time to time. Being prepared and knowing what to look for helps you avoid serious problems. And even when you do all the right things and problems occur, being aware of the steps you need to take can help you avoid trouble in the future.

In this chapter, I introduce you to some of the more likely health problems and ailments that may crop up in your herd and what you do to resolve them.

Managing the Creepy-Crawlies

Regardless of how well you care for your goats, they will get *parasites* — organisms that live and feed on another organism. All mammals have parasites, which isn't completely bad, because their presence stimulates the animals' immune systems and keeps them healthy. But when a goat gets overwhelmed by parasites, trouble starts.

Luckily for goat owners, many of these little critters are host-specific, which means that goats can get them but people can't. By the same token, goats can't get most of our parasites, either. Exceptions are ticks, fleas, and some mites.

Parasites are of two types: *External parasites* live on the outside the body, and *internal parasites* live inside the body, usually in the digestive system.

Controlling external parasites

External parasites are often just an annoyance, but they can lead to bacterial infections of the skin. You can prevent such infections by routine grooming, avoiding indoor overcrowding of your goats, and identifying and treating infestations early.

Lice

Goats can get two kinds of lice: sucking lice and biting lice. Biting lice eat dead skin cells on the goats and make them itch. Sucking lice are more serious — they not only cause itching, but they suck the goats' blood, which can lead to anemia.

Lice tend to take up residence on a goat in winter months. You can usually tell that a goat has lice because it shows signs of itching. Its coat may begin to look rough, and the goat will rub on fences (more than usual), have dandruff, lose patches of hair, and chew on itself.

You can see the lice or their grayish eggs (called *nits*) by inspecting the top of the goat's back with a magnifying glass. You need a microscope to determine whether you're dealing with sucking or biting lice. Sucking lice have large heads, and biting lice have small heads. They are very creepy-looking and have small claws. You can see the differences between sucking and biting lice in Figure 11-1.

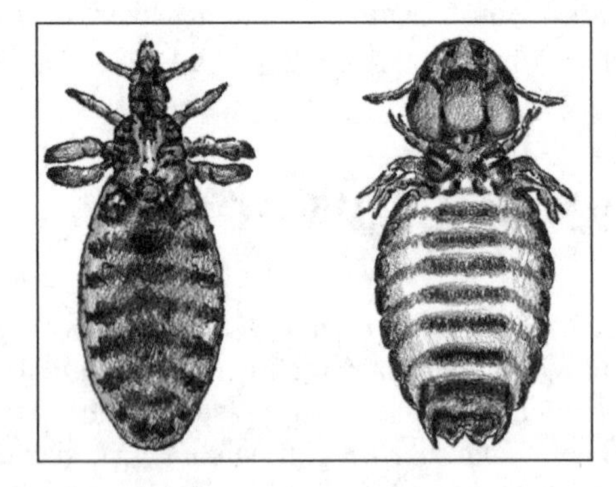

FIGURE 11-1: The head of a biting louse (left) is smaller than that of a sucking louse (right).

TIP

You can often control lice just by regularly brushing your goats or clipping them when the weather is warmer. (See Chapter 9 for more on brushing and clipping.) If the goats aren't severely infested — experiencing intense itching and hair loss — the lice will leave on their own, even without grooming, when the weather is warm and goats spend more time in the sun.

For more severe infestations, treat goats twice over a two-week period, using an insecticide dust such as Co-Ral or a pour-on such as UltraBoss. The labeled use is 1.5 ml per 50 pounds poured along the back and down the face. Do not use this pour-on on a goat more than every two weeks and do not use the dust on lactating dairy goats.

Mites

Like lice, mites infest goats mainly during colder months. They come in two types: burrowing and non-burrowing. The non-burrowing mites usually start in hairy areas of the body, such as the tail, and then work their way along the body. They attach to the skin and puncture it, releasing body fluid. You may see crusty patches and hair loss on a goat under attack from mites.

Some mites live in goats' ears. These ear mites more commonly cause problems in LaManchas because their small external ears are not as protective as the longer ears. Use mineral oil in the ears to smother mites there.

Another type of mites is fond of the scrotum and the area between the legs and belly. I have had success treating these by smothering them with Vaseline.

Burrowing mites are related to the mite that causes scabies in humans. They start in areas that are hairless or have little hair, such as the goat's face or ears. They cause itching and bare burrows in the skin and eventually may lead to thickened skin and extensive hair loss.

You can get rid of both burrowing and non-burrowing mites with *subcutaneous* (under the skin) injections of ivermectin. For best results, treat the whole herd and do a second treatment to ensure that all eggs that hatch after the initial treatment are dead. Continue to check the goats and contact a veterinarian for severe cases or where treatment is not effective.

Fleas and keds

Fleas and *keds* (also called louse flies) are wingless, jumping bugs that can infest goats, usually in the spring or summer. All of them are bloodsuckers, but they usually are more of a pest than a serious health problem. Goats can get fleas from dogs and cats, and they can get keds from sheep.

If your goats have fleas or keds, you probably notice them rubbing, scratching, and chewing, and you will be able to see the bugs upon inspection. You can treat the goat with one of the products that work on lice, such as Co-Ral dust or UltraBoss Pour-on.

Ticks

If your goats are pastured in or near woods, they're a target for ticks. Ticks can be more than just pests because they can spread Lyme disease, Rocky Mountain spotted fever, ehrlichiosis, and other diseases that affect goats and humans. Ticks burrow into the goats' skin, so make sure to remove them as soon as you see them. A tick that has attached to a goat looks like a skin tag and can be as big as a pencil eraser.

TIP

To remove a tick, grasp it with tweezers as close as possible to the head or mouth and pull gently until it lets go. Drop the tick into a jar of alcohol to kill and preserve it in case you want to have it examined later for disease. To prevent future ticks, you can treat the goat with a pour-on or spray that contains the natural insecticide permethrin, such as Ectiban EC.

Minimizing internal parasites

Internal parasites are one of the most common problems of goats, especially in warm, wet weather and climates. They mainly affect the goats' digestive systems, although a few migrate to other parts of the body. All mammals normally have a low level of certain internal parasites — only when the parasites get out of control do they cause problems.

Coccidiosis

Coccidia are parasitic protozoa that tend to be a problem in kids or older, weakened goats. These single-celled organisms are always in the goat's environment and are normally carried by all goats. Only when they reproduce and overwhelm a goat that isn't resistant to them do they become a problem. An overgrowth of coccidia in a goat's digestive tract is called *coccidiosis.*

Kids under the age of six months are at highest risk of coccidiosis because their immune systems aren't yet developed enough to ward off the disease and because their mothers are likely to be shedding a high number of eggs after kidding. These eggs attack the intestinal lining, causing pain and blood loss. The main sign of coccidiosis is diarrhea (called *scours* in reference to livestock) although in some cases an affected goat becomes constipated and dies.

TIP

When an adult goat suddenly dies for no apparent reason, have it checked by a veterinarian for coccidiosis and treat the whole herd if coccidiosis is found.

Some goats die very quickly from coccidiosis; a smaller number of kids have no signs other than failure to thrive. They don't gain weight and are small and thin but don't seem sick otherwise. If you are bottle-feeding kids, you can give Deccox, a drug that is approved in goats for preventing coccidiosis. Save-a-Kid milk replacer contains Deccox, or you can add powdered Deccox to milk. Hoegger Goat Supply sells Di-Methox 40 percent, which they recommend adding to kids' milk to prevent coccidiosis.

If you don't bottle-feed and don't want to give your goats medication unless they're sick, you can help prevent coccidiosis by following the recommendations in the upcoming section "Helping keep parasite problems at bay" and by dealing with diarrhea as soon as you notice it. (For more on diarrhea, see the upcoming section "Scours.")

WARNING

Diarrhea can be a serious problem in kids, dehydrating them fast. If you have a kid that has diarrhea for more than a day, loss of appetite, and weakness, contact a veterinarian and get a fecal analysis to determine the cause. Avoid using Corid (amprolium) to treat coccidiosis because it can cause polioencephalomalacia (thiamine deficiency) in goats. (See more on polioencephalomalacia in the upcoming section "Examining Feed-Related Problems.")

Common worms

The most common worms that affect goats are those that live in the goat's stomach and small intestine. The worm that causes the biggest problem, particularly in rainy, warm areas such as the southern United States, is *Haemonchus contortus,* or the barber pole worm. It is red and white striped, and it sucks the goats' blood and reproduces rapidly.

Anemia is the most common symptom produced by the barber pole worm. (See upcoming "FAMACHA: It's in the eyes" section for information about monitoring for barber pole worm.) You can also identify it by the eggs in a fecal analysis.

Barber pole worm can cause *bottle jaw,* a swelling below the lower jaw. The worm is resistant to many wormers because they were overused trying to combat it over the years. Effective treatment depends on which dewormer works in your area. (See the upcoming "Deworming" section.) Another method for controlling these worms is to give copper orally (see the "Copper deficiency" section, later in this chapter).

Other worms that may take up residence in your goats include the following:

- **Brown stomach worm and bankrupt worm:** More common in fall and winter, these stomach worms can cause diarrhea, rough coat, and thinness and inability to gain weight. Treatment of these worms depends on dewormer resistance. (See the upcoming "Deworming" section.)
- **Tapeworms:** Tapeworms are easy to identify without a microscope because they drop off white sections about the size of a grain of rice in the feces. They cause young goats to get pot-bellied and to develop poorly because the parasites absorb their food. They can also cause diarrhea. A cold freeze can stop the tapeworm cycle in a pasture, but otherwise they can survive in the ground for a year. Treat tapeworms with Valbazen. (Valbazen can cause birth defects if you give it to does in the first 30 days of pregnancy. Many people completely avoid it in pregnancy.)
- **Meningeal worm:** The meningeal worm is spread by deer, which is why you should try to keep deer away from goats. This worm is more common in the fall and winter and needs wet weather. Part of its reproductive cycle occurs in snails and slugs.

  Rather than causing diarrhea, the meningeal worm causes neurological problems in goats, including partial paralysis, circling, blindness, and difficulty walking. If your goat develops these symptoms, contact a veterinarian to determine whether it is meningeal worm and to treat it. Treatment may not be effective and requires not only high doses of dewormers but steroids.
- **Liver fluke:** True to its name, this fluke invades the liver, where it causes internal bleeding and consequent anemia. These parasites use a certain type of snail in their reproductive cycle and affect goats in the winter and spring. In severe cases, the goat will lose its appetite, lie down and not get up, and ultimately die. Less severe cases can cause thinness, rough coat, rapid heart rate, and bottle jaw. The only dewormer that is effective against all stages of liver fluke is Chlorsulan. Valbazen can be used to treat mature liver flukes.
- **Lungworms:** Lungworms are cool-weather parasites. Some use snails and slugs in their reproduction, and others are coughed up by infected goats or are released in the feces and then mature. Hot weather and freezes kill them.

  Lungworms can cause painful breathing, chronic cough, failure to gain weight, and death. When you have a goat with a chronic cough and no fever or other signs of pneumonia, consider lungworms. They can be definitively diagnosed only with a Baermann test, a special fecal exam performed by a veterinary lab.

Testing for parasites

In order to properly use dewormers in your goats, you first have to determine whether your goats have a problem with internal parasites and, if so, what parasites they have, especially if the goats are not showing any signs of illness. Doing a routine fecal analysis before a goat shows signs helps you identify and prevent potential problems before they occur. (See Chapter 10 for more on fecal analysis.)

Giving your goats dewormers routinely and randomly makes parasites resistant to these products, rendering them ineffective.

FAMACHA: It's in the eyes

FAMACHA is a method for evaluating whether a goat is becoming anemic from a parasite overload, most often the barber pole worm. FAMACHA, which requires training, involves looking at a goat's lower eyelid and comparing it to a laminated card with five different colors. The lighter the color, the more anemic the goat is.

FAMACHA enables you to identify and treat only those goats that have a serious parasite problem, rather than treating healthy goats. You can get more information on this method by contacting the American Consortium for Small Ruminant Parasite Control (ACSRCP) at `www.wormx.info`.

Deworming

Sooner or later your goats will develop a problem with parasites, or you'll want to preventively deworm a doe after kidding or a new addition to the herd while it's still in quarantine. You figure out the right dose by first weighing your goat. Then read the label, which tells you the cattle dose by weight. Double the cattle dose for all dewormers except Levamisole (Tramisol), which is one and half times the cattle dose.

If possible, amp up the dewormer's effectiveness by withholding feed for 12 hours before giving the dewormer. Give the dewormer orally by putting it over the tongue into the back of the throat with a drench gun or the tip of the deworming applicator.

Because worms have become more resistant to dewormers, veterinarians now recommend giving two different dewormers from different classes, one after the other, to treat goats with a worm overload. There are three classes of dewormers:

- **Benzimidazoles:** These include fenbendazole and albendazole.
- **Macrocyclic lactones:** These include ivermectin and moxidectin.
- **Imidazothiazoles:** This includes levamisole.

Do not use albendazole (Valbazen) if the goat has been bred or during the first 30 days of pregnancy; it can cause birth defects.

Helping keep parasite problems at bay

Prevent parasite problems by taking a few simple steps:

- Avoid having too many goats in too small an area.
- Don't allow goats to overgraze.
- Regularly clean the goats' living quarters of dirty bedding.
- Don't feed your goats on the ground.
- Keep deer, which can be carriers, away from goats.
- Introduce ducks, guineas, or geese, which eat the snails and slugs that spread certain parasites.
- Deworm goats in quarantine before introducing them to your herd.
- Perform routine fecal analysis (see Chapter 10).

Acquainting Yourself with Goat Viruses and Infections

Viruses and infections are common in goats. Goats are more prone to these diseases when they are under stress, during weather extremes or changes, and when they are overcrowded.

You can avoid some of these conditions by ensuring that goats are tested and found to be negative for the condition before you bring them into your herd. Others can be treated with antibiotics when you catch them early enough. (See Chapter 10 to find out about testing and about identifying illness in your goats.)

Caprine arthritis encephalitis virus (CAEV)

Caprine arthritis encephalitis virus (CAEV), also referred to as a small ruminant lentivirus (SRLV), is a member of the same virus family as ovine progressive pneumonia virus (OPPV) in sheep and HIV in humans. It was named for two of the most common forms of the disease — *arthritis* (inflammation of the joints), which usually is in the form of swollen knees, and *encephalitis* (inflammation of the brain), which shows up as neurological problems. CAEV also can cause chronic mastitis, pneumonia, and weight loss. Both sheep and goats can acquire CAEV or OPPV. In most cases, goats have no symptoms at all but are still carriers.

Researchers don't completely understand how CAEV is transmitted, but it is most commonly spread through body fluids — in the case of goats, colostrum, milk, and blood. Goats that live with an infected goat can also get it, and in some cases CAEV is believed to be passed in utero. CAEV is currently incurable. CAEV is diagnosed by a blood test that indicates whether a goat has antibodies to the virus.

TIP

To prevent CAEV in your herd, know who you are getting your goats from and insist that any new goats or their parents have tested negative for the virus. Have your goats tested the first year after you get them, and if they ever leave your farm, or new goats or sheep come in, continue to test them annually. Breed them only to CAEV-negative bucks.

WARNING

If you have a goat with CAEV, you must keep it isolated from other goats that are CAEV-negative, or plan to have all of your goats eventually become infected. If possible, do not breed an infected goat to an uninfected buck. (The risk of a negative buck becoming infected by a positive doe is unlikely, but some risk exists.)

If you have a CAEV-positive doe that kids, take these CAEV prevention steps to decrease the kids' risk of contracting the virus:

1. **As soon as the kid is born, put it in a separate box and remove it from its mother.**
2. **Wash the kid with mild soapy water, rinse, and dry it.**

   Dry the kid with a blow dryer or towel, making sure it doesn't get chilled.
3. **Put the kid in an area separate from its mother or other goats that are CAEV-positive.**

   You can put multiple kids in the same area as long as all of them are kept from CAEV-positive goats.
4. **Feed the kid within the first half-hour, or as soon as possible.**

   If you have colostrum that has been heat-treated, or is from a doe that is known to be CAEV-negative, or colostrum from a cow known to be negative for Johne's disease, give the kid an ounce or two at a temperature of about 104° Fahrenheit in a bottle. If you do not have safe colostrum, milk some out of the mother, heat-treat it, and give to the kid as soon as possible.

TIP

To heat-treat colostrum, heat it to between 135° and 140° Fahrenheit in a double boiler and keep at that temperature for one hour. Make sure that the temperature does not go higher than 140° Fahrenheit or the colostrum will thicken too much. A good method is to pour the heated colostrum into a hot metal thermos, put the thermos into a water bath, and monitor the temperature of the water.

After the first feeding, feed the kids only pasteurized whole goat or cow milk, kid or sheep milk replacer, or milk from a doe that is known to be CAEV-negative. Note that the kids may be at risk for Johne's disease drinking even pasteurized cow milk. (Check out the section "Johne's disease" to find out more.)

5. **Test the kids for CAEV beginning at six months of age and separate any that test positive from the CAEV-negative goats.**

Abscesses

An *abscess* is an inflamed collection of pus caused by bacteria. Abscesses often appear as lumps in goats' head and neck region, but they show up in other areas, too.

Infectious abscess

Infectious abscesses are usually caused by a foreign object, such as a splinter or a thorn, lodging under a goat's skin and becoming infected. Injections can also cause abscesses. Sometimes you see a lump that enlarges, or you might just notice the large lump all of a sudden.

Bacteria such as staph and strep populate the abscess as the body mounts a defense. If untreated, the abscess can disappear on its own or, more often, it will continue to grow. The outer wall softens until it bursts, releasing a foul-smelling pus. Often the goat loses hair on the site of the abscess.

You can encourage the abscess to ripen by applying hot compresses, or you can lance it with a sharp scalpel. I usually check frequently and wait for the outer wall to thin out and make lancing easier. Always wear gloves to prevent contamination; use paper towels to absorb the pus and burn them when you are done. Then put warm compresses on the wound several times a day to aid in healing. You can also put some triple antibiotic ointment on the area.

Infectious abscesses are not a big risk to the rest of the herd if they burst, although they can spread bacteria. On the other hand, the type of abscess in the upcoming section does put the other goats at risk.

Caseous lymphadenitis (CLA)

Caseous lymphadenitis (CLA) is a highly contagious disease caused by a bacterium, *Cornybacterium pseudotuberculosis.* These bacteria infect the lymph nodes and cause abscesses both inside and outside the body. When the external abscesses caused by the bacteria burst, CLA can spread among the herd. It can also be spread by body fluids and when an infected goat coughs. The bacteria can live in soil, on barn walls, and on other objects for years. It can be brought into a herd in hay.

Although no cure currently exists, you can vaccinate against the disease. (See Chapter 10 for more about vaccinations.)

If you discover an abscess that contains thick, greenish material, assume that it's CLA, isolate your goat, and contact your veterinarian for further investigation. The vet can aspirate the contents of the abscess and have them tested by a lab.

REMEMBER

You can help avoid CLA in your herd by asking the person you're buying goats from whether they vaccinate or have had it in their herd, and specifically in the animal you are buying or its parents. If you find that you have CLA in your herd, separate or remove that goat from your herd because of the risk to other goats.

Hardware disease

Hardware disease is a life-threatening condition caused when a ruminant swallows metal, such as a nail or wire, usually in their feed. The metal perforates the reticulum when the rumen contracts. (See Chapter 2 for more on the reticulum.) Hardware disease is more likely to occur during pregnancy because the growing uterus puts pressure on that part of the body. The perforated reticulum gets infected and can lead to peritonitis. Symptoms of hardware disease are fever, anorexia, increased heart rate, and depression. The most common signs of advanced hardware disease are difficulty breathing, extended head and neck, and standing or walking with back hunched and elbows out.

Goats rarely get hardware disease, but if they do, they may need surgery to remove the object, although antibiotic treatment can resolve it.

TIP

To prevent hardware disease, use hay with strings rather than wire, don't allow goats access to prior building sites, and remove any metal objects from their reach. Goats more commonly swallow strings, which can cause life-threatening impaction. Always keep things picked up and away from goat lips.

Tetanus

Tetanus, colloquially known as lockjaw, is caused when the bacterium *Clostridium tetani* gets into a wound — especially a puncture wound of the foot or leg. It's known to live in the soil, as well as in the intestinal tracts of horses, humans, and other animals. (Remember being warned as a child not to step on a rusty nail, or you'd get tetanus?)

Besides difficulty with opening the jaw, the disease causes neurological problems such as stiffness, a rigid stance, hypersensitivity to touch, and eventually collapse and seizures. If a goat gets this far into the disease, it normally will die within 24 to 36 hours.

Treatment for tetanus is high doses of antibiotics and tetanus antitoxin, as well as making sure the animal is well hydrated and the local wound is treated. Most goats that become infected do not survive.

TIP

Prevention of tetanus is easy: Vaccinate your goats against it. The CDT vaccination discussed in Chapter 10 contains tetanus toxoid, so if you're routinely giving that vaccine, your goats will be protected. Cleanliness is also important.

Two situations in which tetanus is known to occur are disbudding and castration. For this reason, I always give a shot of 150 to 250 units of tetanus antitoxin right before the procedure.

Johne's disease

Johne's (pronounced YO-nees) disease is an infection of the gastrointestinal system caused by *mycobacterium paratuberculosis,* which is in the same family as tuberculosis and leprosy. Johne's disease is chronic and eventually leads to death. A goat with Johne's may not show signs for years but can still spread the disease to herd mates. Signs of the disease are poor coat, diarrhea, and wasting away. You can confirm that a goat has Johne's by blood or fecal test.

Goats can get the disease from sheep or cattle, or even by drinking contaminated cow milk, but they can also spread it through oral-fecal contact. One infected goat in a herd can eventually spread it to all others and will contaminate the area where they live. The bacteria are somewhat hardy and can survive for months in dirt.

To prevent the spread of Johne's in your herd, either cull the positive goats or separate them and do not breed them. Contact your veterinarian to develop a herd health program.

REMEMBER

If you have had any goats that didn't have parasite problems die by just wasting away, consider testing your herd for Johne's. If you are just getting goats, ask whether they have been tested for Johne's disease and whether the prior herd has had any unusual deaths. The only way to confirm that the goat doesn't have the disease is a negative test. (See Chapter 10 to find out about testing for Johne's disease.)

Listeriosis

Listeriosis is a bacterial infection that goats can get from contaminated sileage, soil, and feces of other mammals. Some goats are carriers and never show signs of the disease. You see it more often in the spring and winter, mainly in goats that are indoors all the time.

Goats with listeriosis lose their appetites, become disoriented and depressed, walk in circles, and have paralyzed faces and drooling. Adult goats are more likely than kids to get listeriosis. It's often mistaken for polioencephalomalacia (see the section "Nutrient imbalances"), so I recommend treating for that disease as well as listeriosis.

Goats with listeriosis are more likely to survive when you catch the disease early and treat it aggressively with steroids and high doses of procaine penicillin every six hours. If your goat has symptoms of listeriosis, contact your veterinarian immediately so that you can begin treatment right away. Like any other illness, you need to make sure that the goat is eating and drinking enough.

WARNING

Listeriosis is contagious to humans and can be passed in milk, so do not drink milk from a goat that has had the disease within the past month. Even pasteurization does not kill listeriosis.

Pinkeye

Pinkeye, also known as infectious keratoconjuncitivitis, occurs commonly in goats. The eye becomes inflamed as a result of dust, wind, bright light, or another irritant, and eventually may get infected. This can be spread through the herd by flies or secretions, sometimes leading to an outbreak in the herd.

Mild cases resolve by themselves, but if it doesn't, you need to irrigate the eye with sterile saline and then apply terramycin eye ointment two or three times a day until it clears up. Always wear gloves when treating, to avoid infecting yourself. If the problem worsens, isolate the goat and contact your veterinarian.

Soremouth

Soremouth, also known as *orf*, is a contagious skin disease that affects goats, sheep, and humans. Soremouth usually start as blisters around the mouth but can also affect the nose, the udder, the eyes, the throat, the tail, or other body parts that have a cut or scratch. In some cases, it can even affect the rumen and lungs. After the blister stage, the lesions turn into pustules and become scabby when they break. These scabs contain the virus, which can stay alive for years, and are contagious.

Because of their developing immune systems, young kids are more susceptible, and if they get soremouth can spread it to their mother's udder. Goats that contract soremouth normally develop a strong immunity and can't be reinfected for a year or longer. Most animals recover easily as the disease runs its course, but they are at risk of a secondary infection or fly infestation that can complicate the disease.

Soremouth is usually extremely infectious, although in some cases, only a few animals are affected. I brought it into my herd with the purchase of an asymptomatic goat, and only 4 out of 20 goats contracted it. I have not encountered it in my herd since.

If your goats get soremouth, expect it to last from a week to a month. The only thing you need to do is keep the area clean and try to keep flies and other insects off of them. (See Chapter 5 to find suggestions for controlling flies.) If a goat develops a secondary bacterial infection, contact your veterinarian about treating with an antibiotic.

In cases where a goat develops soremouth lesions in its mouth or throat, you need to make sure that the goat is able to eat. This can be a big problem in kids that are still nursing. If an unweaned kid develops soremouth lesions of the mouth, make sure to separate it from the dam and bottle-feed, to prevent the dam from getting lesions on her udder.

When my goats got soremouth, I tried treating one goat with tea tree oil and one with triple antibiotic and did nothing to the other two. The course of the disease was the same in all cases. If you decide to treat, make sure to wear gloves, because you can contract the virus.

You can quarantine the animal to prevent spread of the disease, but in mild cases I recommend exposing the other goats so they develop immunity. As with other diseases, prevention is the best policy, so ask sellers whether any goats in their herd have soremouth. Another option is to consult your veterinarian about vaccinating your goats against soremouth.

Pneumonia

Pneumonia is an inflammation of the lungs caused by parasites, CAEV, CLA, a sudden change in weather, viruses, poor nutrition, the stress of transport, or poor ventilation. Healthy goats normally have some bacteria in their lungs but have antibodies to protect them. Newborn kids can be prone to pneumonia because they are still developing immunity, but colostrum provides some antibodies.

REMEMBER

The most important things you can do to prevent pneumonia are to provide your goats a clean, uncrowded, and well-ventilated environment; make sure they are not stressed; and watch for other problems (such as a virus or lungworms) that can impair the lungs.

Kids that get pneumonia may exhibit the following symptoms:

- » Coughing
- » Lethargy
- » Moderate fever
- » Rapid breathing
- » Refusal to nurse
- » Runny nose
- » Weight loss

If you don't notice the pneumonia right away, kids' lungs can be damaged. Even kids that recover may be more prone to further pneumonia, develop a chronic cough, or not grow well.

Signs of pneumonia in adult goats include

- » A temperature of 104 to 107° Fahrenheit
- » A moist, painful cough
- » Difficulty breathing
- » Discharge from the nose or eyes
- » Loss of appetite
- » Depression (dullness, isolation from the herd, and indifference to being handled)

If you have a goat with signs of pneumonia, make sure that it is getting enough water. Just like in humans, the adage "Drink water and get plenty of rest" is good advice. You may need to tube-feed a kid (see Chapter 13), or have your veterinarian put an IV in an adult.

Work with your veterinarian to determine whether an antibiotic or pain medication is needed and what is appropriate. Some goat owners, particularly those with large herds, vaccinate their goats against pneumonia caused by two different bacteria, *Pasteurella multocida* or *Mannheimia haemolytica.*

Wounds

Goats often get minor wounds on their faces and, in does, on the udder. Goats with horns are more likely to give or receive more major wounds, and bucks are most likely because they often fight with each other.

To prevent infection, clean a minor wound with warm, soapy water and treat with a spray-on antiseptic such as Blu-Kote or a 7 percent iodine mixture. Check the wound regularly to ensure healing and repeat the treatment, if necessary. Make sure the goat is current on its tetanus vaccination.

Promptly contact a veterinarian if your goat suffers a major wound or a broken bone.

Ringworm

Ringworm is a fungus that leaves a circular, crusty patch of hair loss, most frequently on a goat's head or neck. It's contagious to other goats and to humans and is sometimes spread by cats that have no symptoms. It usually occurs in the winter, when goats spend more time indoors.

Sunlight often resolves ringworm, but if you want to treat it, use gloves and wash the affected area with warm, soapy water, rinse carefully, and apply an antifungal cream, such as Clotrimazole. In severe cases, isolate the goat until the ringworm clears up so it doesn't spread throughout the herd.

Foot rot

Foot rot is a bacterial infection that occurs when a goat spends a lot of time in a wet, muddy area with untrimmed hooves. It causes lameness, foul-smelling feet, and black areas where the hooves join the foot.

You can prevent foot rot by keeping your goats' hooves trimmed and giving them a dry area to live in. If a goat does get foot rot, move him to a dry area, and gradually trim down overgrown hooves. (Chapter 9 tells you about trimming hooves.) Treat the goat with copper or zinc sulfate, such as Koppertox or Dr. Naylor Hoof n' Heel, by pouring it on or soaking the feet daily until the hooves are back to normal.

Rabies

Rabies is not common in goats, but from time to time, there are cases in the United States. Vaccinations are recommended by veterinarians for goats in New Jersey, where there are usually more than 300 cases each year.

Rabies is caused by a virus that affects the central nervous system. The virus gradually affects the animal's brain, which leads to neurological signs such as excessive crying, blindness, circling, depression, and paralysis of the back legs. The animal may be nervous or may even become aggressive (in the furious form of rabies) and then eventually will go down, have seizures, and most likely will die. According to reports from New Jersey, the furious form of rabies is more common in goats.

WARNING

If you suspect that a goat has rabies because of any of these signs or being attacked by another animal, contact your veterinarian. Rabies can be transmitted to humans. A six-month quarantine in a secure facility may be possible, but euthanasia is usually recommended.

Examining Feed-Related Problems

Making sure your goats have proper nutrition is one of the most important things you can do as a goat owner. Many health problems can be traced back to what your goats ate or didn't eat.

Scours

Scours is the term that livestock owners use to talk about diarrhea in their animals. Scours is one of the most common problems in kids as they adjust to food. Too much milk at one feeding or too much grain can cause scours in kids. So can coccidiosis, enterotoxemia, a bacterial infection such as *E. coli* or salmonella, or even pneumonia. Adults can also get scours as a symptom of another illness or, more frequently, in response to too much grain.

REMEMBER

You can help prevent scours by keeping the goats in a clean environment, increasing milk or feed gradually, and making sure the goats get plenty of clean water, exercise, and fresh air.

At the first sign of scours, do the following:

- Stop milk and grain.
- If scours are severe, give some kaolin pectin. In mild cases, this is all you need to get the goat back to normal.
- Make sure that the goat is drinking plenty of water.
- If a goat seems weak, give her some commercial or homemade electrolytes. (See sidebar "Making homemade electrolytes.") Always wait at least an hour before giving milk to a kid who has gotten electrolytes, and never mix the two.

WARNING

MAKING HOMEMADE ELECTROLYTES

You can have a goat that's dehydrated from diarrhea drink this solution; if he is too debilitated, give the solution in a feeding tube. This mixture gives the goat energy and helps keep him hydrated.

You need

- 1 teaspoon salt
- ¾ teaspoon lite salt
- 1 teaspoon baking soda
- 4 ounces corn syrup
- Four pints warm water

Mix all ingredients together. Give a large dairy goat one pint; give a kid or smaller goat ½ to 1 cup every six hours until she regains her energy and diarrhea stops.

Contact a veterinarian and ask for a fecal test if

- » Scours continue for several days despite treatment.
- » Scours are frequent and completely liquid.
- » Scours are bloody or black.
- » The goat is weak.
- » The goat refuses to eat.
- » The goat exhibits signs of pain such as teeth grinding.

Bloat

Bloat is caused when too much gas is trapped in the rumen. Gas is created during digestion, but normally a goat can belch to release it. When the goat can't belch and too much gas builds up, she can die.

Goats are most likely to become bloated in the springtime, when they first have access to lush pasture. They can also become bloated from overeating grain or from getting something stuck in their throat that blocks the release of gas. Alfalfa and clovers put goats at a higher risk for bloat.

REMEMBER

If your goat's left side is bulging, he's lethargic, not eating, and grinding his teeth (a sign of pain), he may have bloat. In severe cases a goat lies down and doesn't want to stand up.

If you have a bloated goat, follow this advice:

- If the bloat is caused by grasses
  - Using a drenching syringe, give the goat a cup of mineral or corn oil.
  - Offer baking soda *free choice* (left out in a container so he can eat it whenever he wants).
- If the bloat is caused by grain
  - Using a drenching syringe, give the goat milk of magnesia. (Give 15 ml per 60 pounds.)
  - Remove drinking water until the problem has cleared up.
- For either type of bloat
  - Walk the goat.
  - Massage the goat's left side.
  - Situate the goat so that his front legs are higher than his back legs. (You can have him stand with his front legs on a step.)
  - Offer the goat roughage such as straw, tree branches, or blackberry leaves when he's ready to eat again.

Enterotoxemia

Enterotoxemia is also called *overeating disease* because it comes about when a goat eats too much grain, lush grasses, or milk. This slows digestion and the intestine becomes poisoned by *Clostridium perfringens* type C or D bacteria, which normally live there. Enterotoxemia occurs more often in kids but can also affect adults — usually when they're under stress, such as from kidding.

In some cases, an affected goat shows no symptoms and then will *go down* (lie down and be unable to get up) and never recover. More often the goat develops a high temperature, severe abdominal pain, watery diarrhea, and loss of interest in food and other goats. Kids with enterotoxemia often cry loudly. In some cases, the goat has seizures or throws its head back. Treatment is generally ineffective. Sadly, only a necropsy after the goat dies can definitely tell you that the goat had enterotoxemia.

Adult goats are more likely to have intestinal bleeding and die from enterotoxemia — sometimes within 24 hours. They can also get an intermittent relapsing, and chronic forms of the disease. Both the intermittent relapsing and the chronic enterotoxemia cause a loss of appetite and diarrhea, but treatment is more likely to be effective.

If you have a goat with symptoms of enterotoxemia

- Think about what he ate recently. If the goat recently got into grain or you increased the amount of grain fed, you may be dealing with enterotoxemia.
- Stop feeding grain and/or milk. Both of these feeds only add to the problem.
- Make sure that the goat is hydrated. Dehydration can kill the goat.
- Feed roughage such as straw or branches. Roughage helps the rumen start working again.
- Give the goat CDT antitoxin as soon as possible. If you give the antitoxin and the goat improves, it is likely to have enterotoxemia.

Because treatment usually comes too late, prevention is important. Keep the goats' area clean and avoid a sudden introduction of or increase in milk or feed. Make sure to keep grain and other concentrates in an area that the goats cannot reach. Goats are smart and sneaky; if they know where the food is, they'll try to get to it. Most veterinarians recommend vaccinating with CDT toxoid. (See Chapter 10 for more about vaccinations.)

Nutrient imbalances

Nutrition-related problems are often caused by improper feeding — mainly when goats get too little or too much of certain trace minerals. You can't always know how much of these minerals are in certain feed or how certain feeds will react with others, but you can follow certain guidelines. (See Chapter 6 for more about feeding.)

Copper deficiency

Copper deficiency is caused when the level of copper in goats' feed is too low, or when too much of another mineral — such as zinc, molybdenum, iron, or sulfur — in feed prevents them from using the copper that's available. Unfortunately, the only way to definitively determine whether a goat's copper level is adequate is on necropsy.

Signs of copper deficiency include a rough, faded coat, light-colored hair around the eyes, bald tail tips, immune deficiencies, mastitis, and myriad other problems. The first step in prevention is to use a proper mineral block.

WARNING

Never use a combined sheep and goat mineral or a sheep mineral because they do not contain adequate copper for a goat.

Talk to breeders and veterinarians in your area to determine whether they have had problems with copper deficiency and how they have resolved it. If you continue to see signs that you believe might be caused by copper deficiency, work with a veterinarian or veterinary school to develop a solution for your herd. This often entails giving a copper bolus (which you can purchase online from Santa Cruz Animal Health [www.scahealth.com] routinely. These can be given by putting the whole capsule in the back of the throat with a pill giver or sprinkling it on their grain. An added benefit is that the copper wires help prevent an overload of barber pole worms (discussed earlier in this chapter).

WARNING

Work with a veterinarian or experienced goat owner to develop a plan for giving copper. Too much can be toxic to goats.

Hypocalcemia (milk fever) and ketosis

Hypocalcemia, sometimes called milk fever, is a deficiency of calcium that some does get during the last month or two of pregnancy, or right after giving birth. It occurs because they aren't getting the right balance of calcium and phosphorus in their feed and are trying to support multiple fetuses or milk heavily.

A doe that develops hypocalcemia gets weak, loses her appetite, stops drinking water, and eventually goes down and sometimes dies. Often the doe aborts her kids, which eradicates the hypocalcemia.

WARNING

Ketosis often goes hand-in-hand with hypocalcemia. (See Chapter 12 for more about hypocalcemia and ketosis.) In ketosis, the body burns its own fat for energy, releasing *ketone bodies* (a by-product created by the breakdown of fatty acids) into the blood. A doe with hypocalcemia gets ketosis because she has stopped eating but still needs energy to grow the kids. A doe with ketosis goes down and is unable to eat, drink, or kid normally. Getting medical attention from a vet is urgent; otherwise, the doe and her kids are likely to die.

You can help prevent hypocalcemia by feeding alfalfa and grain to goats that are in their last two months of pregnancy and are lactating. Grass hay, which is adequate for the other goats, is often not enough for the demands of pregnancy, especially when a doe has multiple kids.

If a doe shows signs of hypocalcemia, the first step is to give her Nutridrench or oral propylene glycol twice a day and make sure she is getting enough to drink; doing so reverses or prevents ketosis. Then contact your veterinarian, who can give her CMPK (a combination of calcium, magnesium, phosphorus, and potassium) to help get her system temporarily back to normal. The vet may also need to give the doe intravenous fluids. The most important step in treatment is to correct her diet, making sure that she is getting plenty of alfalfa and some grain (start slow and increase gradually if she has stopped eating).

Iodine deficiency

Goats normally get iodine from their mineral blocks, but eating too much of certain kinds of plants — including broccoli, kale, turnips, cabbage, and soybeans — can cause them to become iodine-deficient.

An enlarged thyroid gland, or goiter, is a sign of this condition. (The thyroid gland is in the goat's neck just under the Adam's apple.) Iodine deficiency also can lead to problems giving birth, and kids that are born to an iron-deficient doe have short or fuzzy hair and large goiters.

Prevention includes making sure your goats have adequate iodine in their minerals and not feeding them large amounts of plants that can lead to iodine deficiency. Some people paint iodine on the bare area around the tail to treat the goiter. The only other treatment is to increase the iodine in their minerals and make sure they don't eat problem plants.

Polioencephalomalacia

Polioencephalomalacia (goat polio) is a thiamine deficiency that causes central nervous system disturbances, including brain inflammation. Goat polio is more common in a crowded herd. Bracken fern, moldy hay and grain, too much molasses, excess sulfur and the anticoccidial Corid (amprolium) all can cause goat polio, which is more common in the winter and occurs more frequently in kids and young goats.

If a goat has lost its appetite and seems blind, staggers, arches its head back, or stares upward (called *stargazing*), think goat polio. The disease can be confused with listeriosis (which I discuss in the section "Acquainting Yourself with Goat Viruses and Infections" earlier in this chapter). As the disease progresses, the goat may go down, have seizures, and die within 24 to 72 hours.

The only effective treatment for goat polio is administration of high doses of thiamine. High doses of fortified B-complex vitamins given subcutaneously can also help if you can't get thiamine soon enough. Giving thiamine is safe at any time, so don't wait for a lab diagnosis — if the goat responds to treatment and improves, you can conclude that the problem is polioencephalomalacia.

Selenium deficiency

The amount of the mineral selenium in the soil varies depending on where you live. The selenium level in soil affects the hay or other food grown on the land and consequently the goats that eat it, causing selenium deficiency or selenium toxicity.

Selenium works with vitamin E to keep animals healthy. Deficiency in dietary selenium and vitamin E can cause a variety of problems in goats, including muscle weakness, difficulty breeding and kidding, and stillbirths. In kids, it also leads to *white muscle disease,* in which kids become too weak to stand or nurse; sometimes they cough and have trouble swallowing, which can lead to pneumonia.

WARNING

Beware of supplementing selenium without knowing whether your hay comes from a deficient area because you can cause *selenium toxicity.* Goats that have selenium toxicity can become anemic and blind and have neurological problems such as staggering or paralysis.

Consult your county extension office to determine whether your area, or the area where your hay is grown, is deficient in selenium. If it is, work with a vet to put your goats on a program of supplementation, usually injections of BoSe (vitamin E and selenium) twice a year — a month before breeding season and then in the month before kidding in does. You can also keep BoSe (a prescription drug) on hand and give a kid with symptoms of white muscle disease a subcutaneous injection.

Urinary calculi

Urinary calculi are stones in the urinary system. These stones can affect bucks but usually are a problem in wethers, particularly those that were castrated too early. Any stones that form can't get through the narrow opening of these goats' urethras and so stop the flow of urine, causing death if not promptly treated.

TIP

Urinary stones in goats are often caused by an imbalance in phosphorus or calcium. Phosphorus is in grains, and calcium is in alfalfa, so a simple solution to preventing these stones is to not give either of these to a wether. They can also be caused by a lack of regular water, often in the winter when the water freezes. With a good-quality grass hay, a good mineral block, and plenty of fresh, clean water, wethers have everything they need. If you must give treats to wethers, use sunflower seeds and peanuts (in the shell), fruits, and vegetables. Salted, plain corn chips are an excellent treat because the salt encourages more water consumption.

WARNING

Urinary calculi are a medical emergency. The most obvious signs of urinary calculi are straining and inability to pee. If you notice a male goat stretched out trying to pee, check on him and then contact a vet immediately if he isn't peeing. If you don't deal with the problem right away, the goat will ultimately die a painful death as the toxins overcome his body or his bladder bursts.

Poisoning

Goats can be poisoned by a variety of substances, most often plants. (For a list of common poisonous plants, see Chapter 4.) Even if you've rid your backyard or pasture of poisonous plants, your goats can be poisoned by eating plants during an escape or when an unwitting person decides to give them some dangerous plants that they pruned or harvested. This most often happens with rhododendron or azaleas.

Certain plants that aren't normally poisonous can cause nitrate poisoning if they've had too much fertilizer. These include oats, beets, alfalfa, some types of grasses and corn, and others.

Another kind of poisoning is *prussic acid poisoning,* which is caused by cyanide-producing plants such as cherry, corn leaves, certain types of grass, apricot, peach, and white clover. Prussic acid poisoning is more likely to arise from dried or wilted leaves. If you know that a goat was exposed to such a plant and it has bright red mucous membranes, contact your vet.

In severe cases, poisoning leads to death. Ideally, you'll get some warning that a goat has been poisoned through symptoms such as

- Foaming at the mouth
- Vomiting
- Seizures
- Staggering
- Crying
- Difficulty breathing
- Rapid heart rate

Any time one of your goats vomits, suspect poisoning. Goats do not normally vomit and when they do, it's serious and you need to contact your vet. Time is of the essence when you're dealing with poisoning, so don't wait to see whether the goat gets better. Try to figure out what happened and contact your vet right away.

Treatment for poisoning depends upon the poison, and so the first step when you suspect poisoning is to figure out what the goat might have eaten and put it in a place that goats can't get to it. Then call a veterinarian for help.

Always keep an activated charcoal product, such as ToxiBan, on hand. Orally administered activated charcoal is effective in cases where you know that your goat was poisoned by insecticide, rat poison, organophosphate, or alkaloid (found in some plants). Do *not* use it for antifreeze poisoning.

» Dealing with a buck

» Knowing what to expect from goat pregnancy

» Developing a kidding kit

Chapter 12 Breeding and Looking After Pregnant Goats

Breeding season occurs in the fall and is an exciting time of the year that you can feel (and hear) in the air. I can tell you from experience that all of your does will want to be bred. And if a buck is in the vicinity, even if you hadn't planned to breed all of your does, they won't let up for months until you do.

Be prepared to deal with a lot of craziness during this time and stick to your guns about which does you want to breed. Don't let them get your goat!

In this chapter, I cover goat breeding behavior, different ways to get a doe bred, and what to expect when she is pregnant.

Preparing for Breeding

Some goats can breed or be bred when they're as young as two months of age, although the majority are not fertile until four to six months old. This range arises because goats are generally seasonal breeders and don't develop heat cycles until the fall. Doe kids can usually be safely bred at seven months old — which means that they will kid at one year old — unless they are underweight or small in comparison to other does their age.

At the beginning of breeding season you need to do preventive maintenance on goats to ensure that they are in optimal condition and because you may not want to handle the bucks for a while (at least not without gloves and coveralls to protect yourself from the odor, which bucks get from peeing on their legs and faces and from scent glands on the head). This includes

- Giving a BoSe shot, if you're in a selenium-deficient area (see Chapter 11)
- Trimming hooves
- Clipping the belly hair on bucks
- Testing for CAEV (see Chapter 10 for more about testing)
- Doing a fecal analysis and/or deworming

Handling the goats during this time gives you the opportunity to completely examine them to ensure that they don't have any problems that may affect them during breeding season. In some cases, I discovered that a buck was much thinner than he appeared under a bushy coat. This told me that I needed to improve his diet and try to get some pounds on him before putting him through the ordeal of breeding season.

Running through Goat Mating Habits: Courting Is Crucial!

Like many other animals, humans included, goats have their own mating rituals. I wish this book included sound so that you could hear some of the noises that come out of the bucks when they want to get the attention of the does or breed. The pawing, butting, snorting, and blubbering are all essential to breeding. The does play their role, and the bucks play theirs.

Identifying the season for goat love

Goats are mostly seasonal breeders. Some mini and meat breeds can be bred year-round, but in my experience their heat is most pronounced in the fall. Goats that live closer to the equator also may not have a seasonal heat but will ovulate throughout the year.

Beginning in late summer and early fall, the amount of daylight decreases, which signals the does to come into *estrus* (also called *heat*), which ends with *ovulation* (releasing eggs). From as early as July until January, does go into heat about every three weeks. At the beginning of breeding season some does *short-cycle,* which means they come back into heat in a shorter time than three weeks. Normally they cycle every 18 to 24 days thereafter, but if they continue to short cycle, you need to talk with your vet, as your doe may have a cyst on her ovary and won't conceive until she gets a drug to break the cyst.

The decreasing daylight also signals bucks to go into the male version of heat — called *rut.* All they can think of is breeding (sound familiar?), so they stay up late, fight, and sometimes don't eat right.

Manners, or what to expect from your goat

As the days get shorter, you can expect your goats to get more restless because the hormones associated with fertility kick in. Your goats will congregate along the fence line that is closest to the opposite sex and start wearing a path. During this time you need to start planning the breedings to best fit your schedule for kidding five months after the fact.

Does in heat

The heat cycle lasts from a few hours to a few days. You might not be able to tell that a kid is in heat but an adult doe usually creates quite a hullabaloo.

A doe in heat frequently exhibits visible physical changes: Her vulva may swell and become red, and she may have some vaginal discharge. Besides these physical changes, other signs that a doe is in heat are

- **Flagging:** One way to tell if a doe is in heat is vigorous tail-wagging (called *flagging*). I don't know if she is signaling the buck or is just very happy.
- **Vocalizing:** Unless they want something, goats are usually pretty quiet. When a doe is in heat, she may make more noise than usual — from short bleats to longer calls.
- **Parading:** If a doe in heat can see the bucks, she will walk back and forth in his view and rub on the fence more than usual.
- **Acting "buckish":** A doe in heat will mount other goats, or allow them to mount her, and may fight with other goats. These dominance behaviors are caused by hormonal changes that occur during heat. Some people keep a wether (castrated male) with the does as an indicator of heat, because he will also mount a doe at this time.

FLEHMEN REACTION

Especially during breeding season (but at other times as well), you see goats holding their heads in the air and curling back their upper lips. This is called the *Flehmen reaction.* You see it more commonly in bucks, in response to a doe's urine.

The Flehmen reaction causes the goat's nostrils to constrict, pulling the urine odor into the *vomeronasal organ* (an organ consisting of tubes that end in a pair of ducts between the nose and mouth). The vomeronasal organ helps identify whether a doe is in heat by analyzing hormonal products in the urine. Does are believed to use the Flehmen to detect whether bucks are sexually active, bringing them into heat.

I have also witnessed does putting their noses into their kids' urine and flehming. A friend suggested that they were evaluating the health of the kids, and although no studies have investigated this, it makes sense. Other substances can trigger the Flehmen reaction, too, including feces, cigarette smoke, and wine.

- **Decreasing milk production:** A doe in heat may spend so much time trying to attract bucks and fighting with the other does that she isn't as interested in food. Because of this and the hormonal changes she is undergoing, she may temporarily (or even permanently) decrease her milk production.

Bucks in rut

During rut, bucks are ready to breed and can think of nothing else. A buck in rut urinates into his mouth and on his chest, face, and beard, turning them yellow and sometimes causing urine scald with consequent hair loss. The resulting smell, caused by the urine and increased activity in the scent glands near the horns, leads people to think that all goats stink. But that smell also attracts does and brings them into heat.

During rut, bucks blubber, snort, grunt, and *flehm,* or make a distinct face that includes curling the upper lip. (The sidebar "Flehmen reaction" tells you more about this behavior.) They fight other bucks over the does, even if those does are unreachable. They mount each other in a wild display of dominance.

Bucks in rut can mate up to 20 times a day, although doing so puts them under a lot of stress. They often stop eating or eat less during rut, losing weight and body condition.

Supplement your bucks' diet with some grain, leafy branches, and even beet pulp during this taxing time. If you have more than one buck, and particularly if you have more than two, you need to keep an eye on them to make sure that one of them isn't getting injured by the others or sick from lack of food and excessive activity.

Doing the deed

When a doe exhibits signs of heat, you can put her with a buck. He will act very interested and begin pawing and stomping with a front foot. He will stick out his tongue, blubber, and snort. If she urinates, he will put his nose in the stream and flehm. The doe will wag her tail repeatedly and stand still when the buck tries to mount her, although they may circle and go on with this foreplay for five or ten minutes first.

The sex lasts just seconds. I always let a buck breed a doe two or three times in a session to ensure that she is pregnant. If she doesn't come back into heat, you can presume they were successful. If you have a young buck or one that you want to breed more than two does in a day, I recommend only one or two successful matings. Baby bucks have a limited amount of semen available.

In rare cases, even when a doe appears to be in heat, she may dislike a specific buck and refuse to be bred. Sometimes you can put her in a stall with the buck and she comes around, but other times she will do everything she can to get away. Older does often don't like baby bucks, who either don't smell sexy enough yet or are not aggressive enough and don't woo properly.

You can sometimes trick a doe by letting her love a buck she likes through a fence and bring the buck you want to use behind her and let him breed while she is thinking about the other buck. You can hold a doe in place so the buck can breed her, but use your judgment. Like people, goats have preferences and may get overly stressed when forced.

Housing a buck after breeding

If you let your buck live with the does during breeding season, you need to start making plans for housing him when the kids are born. Not only will he be a pest when a doe goes into labor, but the risk of him rebreeding the doe too early is very real. That gives you about five months to get another area set up, if you don't already have one.

If you have only one buck, you can put him in a fenced pasture adjacent to the does'. That way he can see them and interact with them. If you're planning to move him to a more distant area, consider getting a wether or another buck as a companion, so he isn't too lonely.

Finding Breeding Solutions When You Have Only Does

Many goat owners don't keep bucks because they don't have enough does to justify the cost and trouble of keeping one. They need separate quarters and another buck or wether for a friend, and they are very aromatic! And if you have an urban farm, they may not even be allowed by the municipality you live in.

TIP

When you don't have your own buck, you need to pay for *buck service* to get your doe bred. Find out from the breeder what he charges for this service, and if you plan to register the kids born to your doe, you need to remember to get a service memo from the breeder after your doe is bred. A *service memo* is a document signed by the buck owner verifying that his buck bred your doe and the approximate date. Most breeders give a second breeding free if the doe doesn't become pregnant.

In this section, I give you ways to tell whether your goats are in heat and get them bred if you don't have a buck.

The invaluable buck rag

A *buck rag* is a piece of cloth that has been rubbed on the head, neck, chest, or beard of a buck that is in rut and then stored in a jar. These parts of his body are covered with his perfume. A buck rag is a great tool to tell you when a doe is in heat if no buck is around. It usually drives the does crazy!

TIP

If you're a novice goat owner and are having trouble telling when one of your does is in heat, ask another breeder who has bucks if you can come and get a buck rag. Then when you think a doe *might* be in heat, just pull out the buck rag. Although you can't always tell with a young, inexperienced doe, the buck rag usually elicits a noticeable response. The doe starts flagging, getting excited, rubbing on the rag, and vocalizing.

The response you get from a doe that has smelled the buck rag tells you whether it's time to try to bring in a buck, take her on a date, or artificially inseminate her.

Leasing a buck

A common method for getting your does bred without having to keep a buck is to lease one. Leasing a buck may not require any additional space, if he can live with the does during the time required for breeding. If you have more than one breed or does you don't want to breed, you need a separate pen for the leased buck.

I leased a buck for my first two does. The buck, Harley, got to come and live with a couple of hot chicks for about six months — probably the best six months of his life. The girls and I liked him so much that we let him stay almost until the kids were born. (You don't want a buck in with a doe that has recently kidded, because she can get rebred very quickly.)

Talk to local breeders about leasing a buck. Make sure the buck is healthy. (See Chapter 7 to find out about knowing a healthy goat from a sick one.) If you can find an acceptable buck, the price is usually reasonable (a lot less than keeping one). You will have to negotiate the length of time you keep him. Unless you see him breed your goats, you need to have him for at least three weeks, so they can go through a full breeding cycle.

Less commonly, a breeder will have a separate space that can accommodate one or two does, along with the breeder's buck, for a short-term live-in breeding. If you can stand to have your girls gone for a month or so, consider this option.

One-night stands, or driveway breeding

Driveway breeding is the most common method used by goat owners who don't have a buck. Usually you take the doe to the buck.

You need to talk to the buck owner beforehand, so that she will be expecting your call. Then when your doe shows signs of heat or responds to the buck rag, you put a collar and a long leash on her, load her up in your vehicle, and drive her to the farm where the buck lives.

Some buck owners have a separate stall to put the breeding animals in, but generally you just take the goat out of your vehicle and hang on to her on the long leash while the buck owner lets his goat out. When they are done you still have control to end the romance and get her back in your vehicle.

Artificial insemination

Artificial insemination (AI) involves collecting buck semen and transferring it to a doe's reproductive tract. The semen is usually frozen and kept in a nitrogen tank, and then thawed prior to use.

Artificial insemination is a less common method of breeding goats, usually reserved for bucks and does with outstanding genetics. AI enables you to use semen from bucks that aren't available to you and also to use semen from outstanding bucks that are deceased. The initial cost for purchasing the necessary equipment is high. The biggest cost is a liquid nitrogen tank for storing the semen.

If you are interested in this method, talk to other breeders who use it. You may be able to cut your costs by sharing a semen tank with other breeders, or another breeder may let you store some semen in theirs. You can also save by hiring a semen collector to retrieve the semen when she comes to a neighboring farm.

Looking into the Finer Points of Goat Pregnancy

After your doe is bred, you don't need to do much right away. For the first three months you won't even be able to tell that she is pregnant, unless she has multiple kids gestating and starts widening early. If she's still milking from the prior year's kidding, you can continue to milk until she is three months along.

You don't need to change her feed right away, either. As long as she is already getting good nutrition, she will be fine. But when she reaches the three-month mark, you need to adjust her ration. (Chapter 6 tells you about feeding for pregnancy.)

Length of gestation

A goat's *gestation* (the time between breeding and birth) is approximately 150 days, although it can vary between 145 and 155. Nigerian Dwarves frequently kid at only 145 days, and goats that have poor body condition or nutrition often kid later than 150 days.

TIP

Always write on your calendar 145 days after the date that a doe was bred as the due date. You can start checking her ligaments and watching her around this time. (See the upcoming section "Knowing when she will kid [and what to do!]".) You can find a due-date calculator online if you want to save the trouble of trying to calculate the date.

False pregnancy

False pregnancy, or *pseudopregnancy*, is not an uncommon phenomenon in goats (and some other animals). In a false pregnancy, the goat has all the signs of being pregnant, such as an enlarged udder, milk production, and uterine cramping.

During false pregnancy, a pregnancy test will even come out positive because the hormone progesterone is being produced just like in a real pregnancy.

False pregnancy is sometimes linked to uterine infection. It can end at any time but most often goes full term and ends in a *cloudburst*, or the release of fluid but no kid or placenta. The goat will then go back into her normal cycle and can be bred again.

False pregnancy is frustrating for an owner who has waited five months for a goat to kid, but usually it isn't any more than an annoyance.

Dealing with common pregnancy problems

Most pregnancies are uncomplicated. You can help to ensure that problems don't arise by providing proper feed and shelter and paying attention to your herd's health.

The upcoming sections tell you about some of the most common problems that do occur in pregnancy.

Abortion and stillbirth

Abortion is when a pregnancy ends before it reaches full term. When a kid is born dead at term, it is a *stillbirth.* Both of these problems can have a variety of causes, including the following:

- **Malformation or genetic defect:** Kids with genetic defects are usually aborted early in the pregnancy, and often you may not even know the goat was pregnant. These abortions cannot be prevented.
- **Stress:** Stressors such as poor nutrition, cold weather, overcrowding, or poor diet can lead to abortion. You can prevent these stressors by providing proper shelter, not housing too many goats in a small area, and feeding a balanced diet. (Chapter 6 tells you more about feeding during pregnancy.)

  Avoid transporting a pregnant doe or moving her to an unfamiliar place during her last month of pregnancy. The stresses can cause her to abort.
- **Infectious diseases:** Fifty percent of abortions in goats are believed to be caused by infection. The list of infectious diseases that can cause abortion is quite lengthy (see Table 12-1).

  Infectious abortions are a risk for the whole herd, particularly other pregnant goats. If more than one goat aborts, save the fetus and placenta in the refrigerator and call your veterinarian or veterinary school to perform a *necropsy* (examination after death) to determine the cause of death.

TABLE 12-1 Some Infectious Causes of Abortion

Disease Agent	Timing of Abortion
Chlamydiasis	Last trimester
Brucellosis	Late term
Campylobacteriosis	Last six weeks
Border disease	Any time
Listeriosis	Last trimester
Salmonellosis	Mid to late term
Toxoplasmosis	First half
Q fever	At term (stillbirth)

- **Poisoning:** Some plants and medications, such as certain dewormers and steroids, can cause abortions in goats. (See Chapter 4 for information about poison plants.)
- **Injury:** Occasionally a hard butt in the side by another goat will cause a doe to abort.

Hypocalcemia

Hypocalcemia, often called *milk fever*, is a deficiency of calcium in the blood that arises when a doe doesn't get enough calcium in her diet to support her needs and the needs of her unborn kids. It most commonly occurs at the end of pregnancy, but does can also get it during lactation, especially if they are heavy milk producers.

Signs that a goat is developing hypocalcemia are a loss of appetite, particularly for grain, at 12 weeks of pregnancy or later — even after kidding. This loss of appetite leads to progressive weakness, staggering, depression, low temperature, and lethargy. These symptoms are caused because the muscles are not getting the calcium they need to operate. Eventually, a goat with hypocalcemia will go down and not get up and, without treatment, she will die.

You can prevent hypocalcemia by making sure that the doe has an adequate diet during pregnancy and lactation. She needs alfalfa for calcium, and getting too much grain early on can interfere with her getting it. She needs about two parts calcium to one part grain.

The steps you need to take to prevent hypocalcemia are different, depending on whether the goat is lactating (and getting grain and alfalfa) at the time she is bred:

- If she is lactating and getting a heavy ration of grain and alfalfa at the time of breeding, continue this grain ration throughout the pregnancy. Continue giving alfalfa unless you dry her off (stop milking) before she is three months pregnant. If you dry her off, switch to grass hay until the three-month mark.
- If she is bred when dry, do not give grain or alfalfa until the last two months of the pregnancy. Start slowly with only a handful of grain, and gradually increase the amount of grain you give her. Slowly replace her grass hay with alfalfa.

TIP

To feed alfalfa only to does that are pregnant, use alfalfa pellets, which they can eat along with their grain on the milk stand.

To treat hypocalcemia, *immediately* give the doe Nutridrench (according to directions on the bottle) or 60 ml of oral propylene glycol twice a day. This will give her the energy she needs to go on. Then contact your veterinarian to start her on 30 ml CMPK, a prescription combination of calcium, magnesium, phosphorus, and potassium, subcutaneously every two hours to address the mineral imbalance, and provide intravenous fluids, if necessary. You can tell that this is working if her heart rate returns to normal. You may need to continue giving the CMPK on a daily basis until she kids, especially if she develops the same symptoms. (See Chapter 11 for more about hypocalcemia.)

TIP

If the goat is within five days of kidding and doesn't get better right away, ask your veterinarian for a drug called Lutalyse to induce kidding and eliminate the stress of supporting the kids.

Ketosis

Ketosis is a metabolic imbalance that usually goes hand-in-hand with hypocalcemia. It is caused when a goat doesn't get enough energy because she has stopped eating. The body then releases fatty acids that are used by the liver, which produces *ketone bodies* (by-products produced by the breakdown of fatty acids for energy). (See Chapter 11 for more about ketosis.)

Ketosis is more common in does that are overweight at the beginning of pregnancy and in those that have multiple fetuses. A doe with ketosis has sweet-smelling breath in addition to the symptoms of hypocalcemia. Treat ketosis with propylene glycol or Nutridrench, and also treat for hypocalcemia. (See the section "Hypocalcemia.")

Getting Ready for Kidding

To make kidding go more easily and protect both *dam* (mother) and kids, you need to take certain precautions before kidding. In this section, I talk about routine pre-kidding care for the dam and how to prepare a kidding area.

Preparing the doe

During the last two months of pregnancy you need to do some routine care to make sure that the doe and you have an easy kidding. You can take a few simple steps to ensure that everything goes well both during and after the kidding:

- **Give a BoSe shot.** If you are in a selenium-deficient area give (or have your veterinarian give) her a BoSe shot, which is a prescription selenium/vitamin E combination, a few weeks before the expected kidding date. Doing so helps prevent uterine dystocia (abnormal labor), aids in passing the placenta, and helps prevent white muscle disease in kids. (Chapter 13 tells you more about white muscle disease.)
- **Vaccinate with CDT toxoid vaccine.** If you vaccinate your goats, give a CDT booster shot four weeks prior to the expected kidding date. Doing so provides the new kids with some immunity from enterotoxemia and tetanus in the first few months of life. (Chapter 10 tells you more about this vaccine.)
- **Trim the tail and udder area.** I like to trim the tail and udder area a week or so before the expected kidding date to help the doe stay cleaner during and after kidding. For goats with hairy udders, a trim makes it easier for kids to find the nipple and start nursing. A few weeks after kidding, the doe will develop some bloody discharge that builds up around the tail area. Removing the hair that the discharge normally sticks to minimizes the build-up.
- **Stop milking.** If your goat is still lactating, you need to stop milking her for the last six to eight weeks of the pregnancy so that her energy goes into growing the kids and preparing her body for the pregnancy. She won't need as much grain when she stops milking, but don't decrease it suddenly.

Setting up a kidding pen

Start getting a pen ready for kidding a few days before the pregnancy reaches the 145-day mark. Clean any used straw or wood shavings from the pen, sanitize the walls with bleach water (especially if the pen was used previously for a sick goat or for another kidding), and put in a thick layer of fresh straw or wood shavings. Sanitize an empty bucket and have it ready for water when the time is right.

TIP

A baby monitor is a big help for identifying when the time has come for you to go help with kidding. If you have one, set it up when you prepare the kidding pen and put the receiver where you can hear what is going on when you're not in the barn. If your goat is in an early stage of labor you may hear groaning, cries, or nothing at all. If the goat is in the second stage of labor (pushing the kid out) you are likely to hear loader cries, heavier breathing, or *mama talk*, a series of short cries meant only for newborns. Sometimes you will be alerted by the cry of a newborn kid.

Prepare or check your kidding kit to make sure you have all the supplies you need and wait for the goat to kid.

Being prepared with a kidding kit

You can take a lot of the stress out of kidding just by being prepared. That means having a clean, separate kidding area, knowing when your goat is going to kid, and having the supplies that you need for a routine kidding or for a kidding that doesn't go according to plan.

Putting together a kidding kit takes a little time up front, but it saves you time and trouble in the long run. At a minimum, make sure you have the following items available at kidding:

- 7 percent iodine for dipping cords
- A film can or prescription bottle to hold iodine when you dip cords
- Flashlight, if you don't have good lighting in your kidding area
- Phone numbers for
  - Your veterinarian in case of complications that you can't handle
  - A more experienced goat owner for questions about kidding that you think you may be able to handle
- Dental floss for tying cord before cutting
- Bulb suction for clearing kids' nostrils or airway
- Old towels (one for each kid you expect, plus an extra)
- Betadine surgical scrub for washing the goat and your hands if you need to assist with the delivery

- Sterilized surgical scissors for cutting the umbilical cord
- Disposable examination gloves
- K-Y Jelly or obstetrical lube
- A feeding syringe and tube for feeding weak kids (see Chapter 13)
- Empty feed bags to put under goat and to use for after-kidding clean-up
- Empty pop bottle with Pritchard teat, in case you need to bottle-feed

TIP

If you don't have a hot water supply near the kidding area, bring a bucket of warm water for washing up. In a pinch, you can use a mild dish soap for washing your hands and your doe's vulva. You can also use soapy water for lubrication if you have to assist the doe.

Knowing when she'll kid (and what to do!)

As the time nears for your goat to kid, you may get just as nervous as she does. She most likely can kid on her own, but you want to make sure that she has a clean, safe place to do so. I give you a few tips on how to tell when your goat is getting close to kidding.

Reading the ligaments

A goat's rump is normally flat and solid, but as a doe gets to the end of pregnancy, that changes. Her tailbone becomes elevated, and the ligaments that connect it to her pelvis begin to stretch and loosen in preparation for the journey the kid (or kids) will make from her body. Sometimes you can tell that she will kid soon when you see a hollow on either side of the tail.

One of the best ways to identify an impending kidding is to feel the two tail ligaments located on each side of the tail. Feel a doe that isn't pregnant and you will notice that those ligaments are very firm. The same will be true of a doe that is pregnant but not ready to kid.

When these ligaments begin to get soft, and then completely vanish, you know that the goat is due to kid within 24 hours. You may make a mistake the first few times you try to read the ligaments, but over time you find the technique to be almost foolproof. Figure 12-1 shows you how to find the ligaments.

A few weeks before the doe is ready to kid, start feeling her ligaments routinely. One day you will find that they've turned to mush, and then you will know that it's time to put her in the kidding pen.

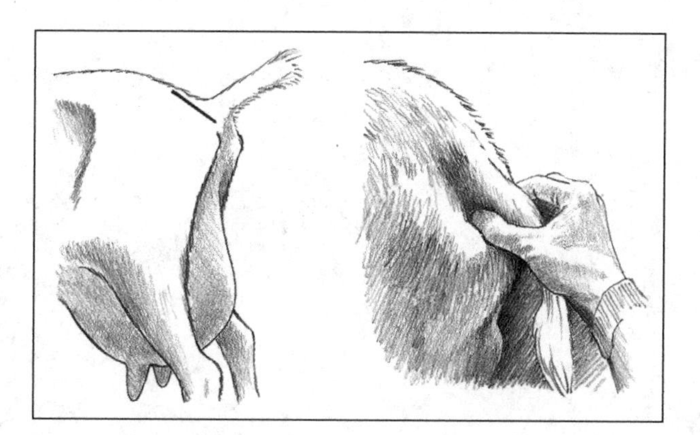

FIGURE 12-1: Check the ligaments on a goat to tell whether she is going to kid.

Identifying other signs of impending kidding

Besides softened ligaments, a doe will show other signs of kidding. Each doe might exhibit different signs, so keep an eye out for a change in behavior. Some other signs to look for include

- **Isolation:** The doe stands off from the crowd, sometimes seeming "spaced out."
- **Mucus discharge:** You may observe some whitish or yellowish discharge on her vulva.
- **Firm, shiny udder:** Her udder may become tight and filled up, called *bagging up.*
- **Loss of appetite:** She may become uninterested in food.
- **Personality change:** She may start fighting with other goats or become overly friendly to you when she was previously standoffish.
- **Restlessness:** She may lie down, then get up, paw at the ground, and just seem uncomfortable.

When you have checked her ligaments and they're soft, or when you notice her exhibiting any or a combination of these signs around her due date, put her in the kidding pen, give her some fresh hay or alfalfa and observe her in this environment.

When you have determined to your satisfaction that this is the day, turn on the baby monitor and leave her to focus on the mysterious process of having a kid. (Chapter 13 tells you about the kidding process.)

» Dealing with newborn kids

» Developing a feeding program

Chapter 13

Now Comes the Fun Part: Kids!

Nothing is cuter than a baby goat. They usually come into this world with only their mother's help, but it doesn't hurt to be prepared for problems, just in case. You also need to know how to help these little guys get off on the best foot and how to help their mother recover.

In this chapter, I address what to expect with kidding, how to deal with those newborn kids, and developing a successful feeding program to help them grow up strong and healthy.

Grasping the Basics of Kidding

For thousands of years, goats gave birth to their kids without any human intervention. Those that had problems died, and their kids also may have died, while the goats that kidded easily survived. When we began to tinker with and "manage" goats, nature no longer ruled and goats that would have died began to survive.

Most goats can give birth without human help, but if you want to raise goats, you need to know the basics of kidding so you can help when you have to.

Unless you discovered your doe after she started active labor, after you get her in the kidding pen, you can leave her to the solitude that most animals prefer during labor. You can listen on a baby monitor and check on her intermittently to see how she's doing as she goes through the stages of labor. When she reaches the second stage of labor, you usually know that it's time to be there because her breathing changes, and she makes loud cries as she is pushing the kid out.

Knowing what to expect from labor and birth

In the *first stage of labor* the uterus contracts and dilates, forcing the unborn kid against the cervix (neck of the uterus). This process usually lasts about 12 hours for goats that are kidding for the first time, but it can be longer or shorter. Every goat is different.

During this stage the goat will be restless, lying down and then getting up, unable to get comfortable. She may look at her side, like she can't figure out what is going on. She also may lick herself, or even you, in anticipation of the kids coming out. Most goats want to be left alone during this time, and their labor may even slow down or stop if people are around. A doe that you're very close to may react the opposite way — wanting you there comforting her. Some goats stand up to deliver their kid, and others lie down. Figure 13-1 shows a doe lying down to kid.

FIGURE 13-1: A doe may lie down to deliver her kid.

The *second stage of labor* is when the doe pushes the babies out of her uterus. Her contractions get stronger and if the kid is lined up correctly, it will start moving down the birth canal.

TIP

If you're listening on a baby monitor when the doe enters second stage, you may hear the doe cry out. (On the other hand, like people, some goats give birth silently.) From the time the goat starts pushing until the first kid is delivered should be only 30 minutes. If it takes longer than this, the kid may be malpositioned or the doe may have another problem. Investigate whether the kid is stuck or coming out wrong to determine whether you or a vet need to intervene.

You will see thicker discharge, sometimes tinged with blood, and then a bubble at the opening of the vagina. This is the amniotic membrane. If you look in the bubble you usually see a nose and one or two little hooves. Figure 13-2 shows you a goat in a normal position for birth.

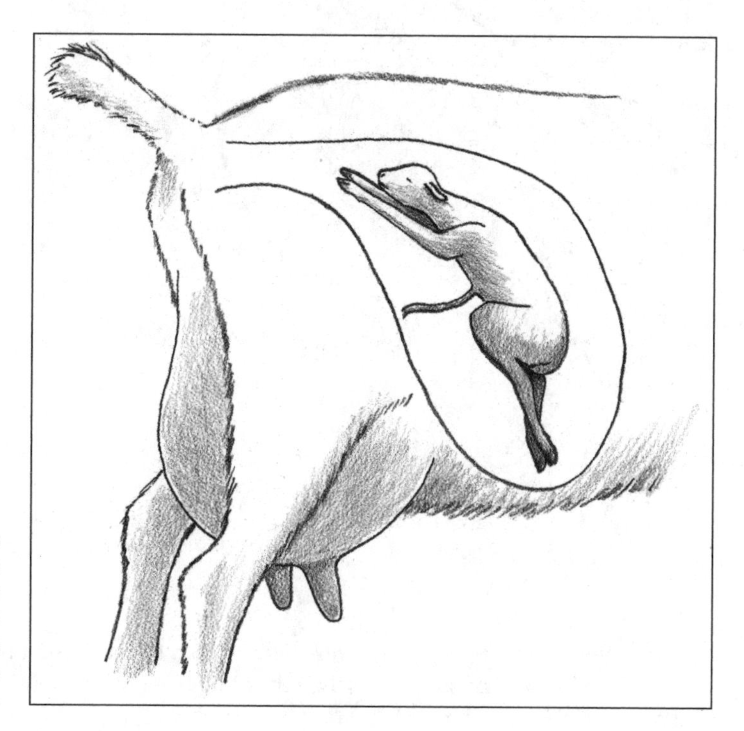

FIGURE 13-2: In a normal birth presentation, the goat is positioned head first, with its hooves outstretched.

After the bubble appears, the doe will continue to gradually push the kid out, sometimes stopping to gather her strength. Sometimes she will circle around, try to get to that bubble, or lick at her sides or your hand, expecting a baby. Within a half hour, the baby will slide out, usually to the accompaniment of a yelling goat. Often kids are born *in the caul* (still in the amniotic membrane). This provides a cushion for an animal that drops a distance to the ground.

If the amniotic membrane doesn't break when the kid comes out, break it and clean the fluids from the kid's mouth and nostrils. The kid should breathe, cough, or shake her head to clear excess mucus. If the kid doesn't start breathing right away, sticks her tongue out, or otherwise has difficulty breathing, swing it as I describe below.

Breech presentation (back feet first; see Figure 13-3) is another version of normal in goats. If kids are small, even a *frank breech presentation* (tail first) doesn't present a problem. The risk in a breech birth is the possibility of inhaling amniotic fluid. A gentle, steady pull on the hind legs in a breech birth will help to ensure that the kid's head comes out promptly.

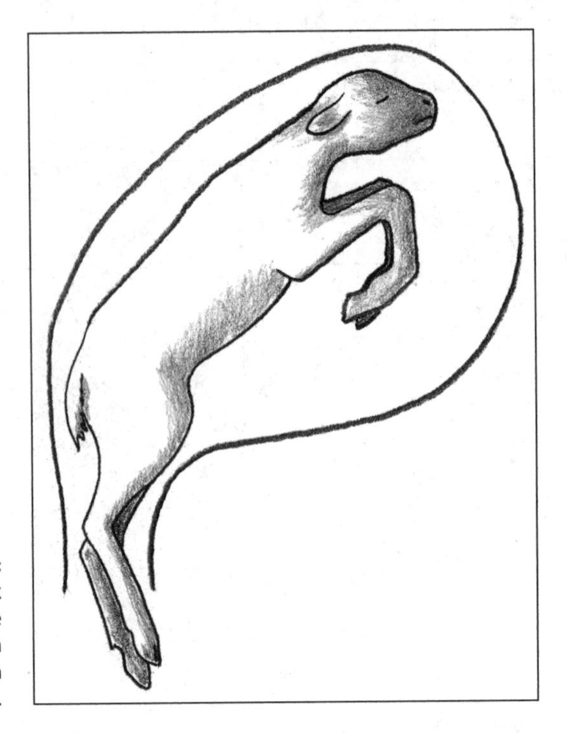

FIGURE 13-3: The kid's back feet are positioned first in a breech birth presentation.

TIP

I recommend swinging kids that were born breech to clean out any fluid they may have inhaled. I used to do it only when they had trouble getting started breathing, but after the death of a breech kid that had seemed all right, I now do it with every breech. To swing a goat kid:

1. **Wrap a towel around the kid (the process can be messy).**
2. **Hold the kid by the feet with one hand and in the area between the head and neck with the other hand.**

3. **Swing it back and forth several times with head facing out to clear the lungs.**

   Make sure you are in an area where you won't hit anything and be aware that the kid is slippery.

4. **Check the kid's breathing and repeat the process if it isn't breathing.**

Another position that can cause a problem is *front legs back.* The birth is stalled and you see a nose (sometimes with a tongue hanging out of the mouth), but no hooves. That kid has its legs back and unless it's very tiny, it won't be getting out without some help.

TIP

When a kid is in the front legs back position, you have to get the legs out so that the kid is in "diving" position, with its front legs stretched out alongside its head. Don't try this unless you have experience or are very comfortable assisting. Your hands must be very clean and well-lubricated and you will need to push the head back into the mother, follow the neck down to the shoulder, locate the leg, carefully draw that hoof up, and bring it out as gently as possible to avoid scraping the uterus or vagina. The hooves are soft at birth, which helps.

Other positions that require assistance and can be difficult to reposition include *transverse* (across the uterus with a side near the cervix), *head back* (hooves out but head back, usually to one side), or even *crown presentation* (the top of the head coming first). The only doe of mine that ever had a *cesarean* (surgical removal of the kid) was one with a kid whose crown was presenting first. Neither the vet nor I could get the nose up to get him out.

If you encounter a kid in the wrong position, call a veterinarian or an experienced goat owner to assist you or talk you through it. If you are unable to get the kid into the correct position you need to have a veterinarian perform a cesarean.

If you plan to bottle-feed the kid or raise it on CAEV prevention (removing kids from the mother immediately and feeding only pasteurized milk or milk replacer; see Chapter 11), take the kid as soon as it's born and before the dam even sees it, because she will start cleaning it up instinctively. If the kid will be dam-raised, then you can help the mom with the clean-up. (See the section "Caring for newborn kids" to find out how.)

Handling multiples

If the doe is carrying more than one kid, you may barely be finished with the first (or second, third, or fourth) one before the next is on its way. Multiples usually are smaller, which means they come out more easily, but they also can get tangled up and be malpositioned, so that you or a veterinarian need to assist.

TIP

Even when they do well during the kidding, kids from births of three or more at a time can get lost in the shuffle and be neglected by a doe or be pushed out by a more assertive littermate. Mom has only two teats and if two of the kids are big, greedy bucks and one is a small doe, guess who will lose. Plan to spend more time and assistance right after kidding to make sure that such kids are getting what they need. In some cases, you will have to bottle-feed the smaller, weaker kids.

Taking Care of Mother and Kids after Kidding

After a kid is born, he needs to be cleaned, undergo a series of minor procedures, and then demonstrate his innate nursing skills. His dam also requires attention after her hard work. In this section, I tell you what to do for mother and baby after a kid is born and how to handle problems with newborn kids.

Caring for the new mother

The first thing to do when a doe is done kidding is to get her a bucket of warm water with a little molasses (about two gallons of water with 1/4-cup molasses) for energy. My goats sometimes drink a whole gallon. Then get her some grain and some fresh alfalfa to munch on while her kids learn to walk and nurse, and while you wait for the placenta.

After a doe has kidded she goes into the third stage of labor: delivery of the placenta. This stage normally can take up to 12 hours, but usually the doe passes her placenta within an hour or two of kidding. If she has not done so within 12 hours, contact your veterinarian. Dispose of the placenta by burying it deep, composting it, or burning it.

The dam will try to eat her placenta — most mammals do. I allow her to take only a few bites because I think that much might be good for her, but I don't want her to choke on it — something that happened to my first goat, Jinx.

Clean the kidding area of soaked straw and feed bags and add fresh straw. If you will be bottle-feeding, milk the doe out (milk all the colostrum out of her udder), heat-treat the colostrum if needed, and feed it to kids or freeze it for emergency. (See the upcoming section "Caring for newborn kids" for how to heat-treat colostrum.)

If you haven't dewormed the doe in the last month, check her FAMACHA score (see Chapter 11) and do a fecal exam within the first few days after kidding. After kidding is the one time that I sometimes deworm a goat without doing a fecal. Because kidding is a time of stress on the doe's body, she's at risk for what is known as a *periparturient egg rise,* in which the number of eggs that parasites are releasing increases. This is even more likely when kids are born in the springtime, when the parasite life cycle starts up again.

As you're watching or helping the kids get their first colostrum, check out the doe's mammary system, looking for problems like lopsidedness, hardness, or heat. Continue to monitor for mastitis over the next week or so. (I tell you more about mastitis in Chapter 15.)

Caring for newborn kids

I love watching the newborn kids try to stand, then go to a wobbly walk and eventually lurch toward their first food. They may complain about being handled, but they need to be warmed and stimulated so they can get up.

After the birth of a single kid, or between the births of multiple kids, take these steps with each newborn:

1. **Use a towel to help the dam clean the kid.**

   Make sure you uncover the face first and determine whether the kid is breathing. If he isn't breathing, rub the body to stimulate him; if that doesn't work, swing him as I describe in the section "Knowing what to expect from labor and birth." Stimulating the kid is critical to getting her to stand up, walk, and nurse.

   If you are on a CAEV prevention program or plan to raise the kids separately on a bottle, don't let the dam clean the kid. Instead wash the kid and put him in a separate box from other newborn kids until each has been washed and dried. (See Chapter 11 for more information on CAEV prevention.)

2. **Tie the cord with dental floss and cut it about an inch from the kid's body.**
3. **Dip the cord.**

   Pour some iodine into a film can or prescription bottle and hold it over the umbilical cord stump up to the belly. Turn the kid to coat the whole cord. Treating the cord with iodine helps prevent navel ill. (See "Dealing with kid problems" to find out about navel ill.)

4. **Check the kid for sex, number of teats, and any abnormalities.**

5. **Feed the kid.**

   Put the kid under the dam if it will be dam-raised. Watch the kid for the next 15 minutes or so. If she has trouble getting latched and sucking well, help her out by moving her near the teat or putting the teat in her mouth. Some kids really resist; other little pigs (er, goats) latch on and don't back down.

   Offer a warm bottle with several ounces of heat-treated colostrum if the kids will be bottle-fed.

TIP

To heat-treat colostrum, heat it in a jar in a hot water bath to 135° Fahrenheit and hold it at that heat for one hour. Do not let the temperature go higher than 140° or below 130°.

The importance of colostrum

Colostrum is the first milk that the doe produces for her kids; it's rich in nutrients and proteins called *immunoglobulins*, which provide immunity against disease, and it stimulates the kids to expel *meconium* (the first feces) from their intestines.

One of the most important things to do for newborn kids is make sure that they get plenty of colostrum soon after birth. Ideally, they get their first colostrum within 30 minutes after birth. If you can't do it in that time frame, do so as soon as possible. (Sometimes you have to heat-treat colostrum first, which takes about an hour; see the section "Caring for newborn kids.") Kids are able to get the immunoglobulins only in the first few hours after birth.

WARNING

Use one of the dried colostrum alternatives that you can buy online or at a feed store only when you have no other alternative. Most don't provide kids with the immunity they need. If you have to resort to this, it's important get a product made from bovine Immunoglobulin G (IgG). The ideal is goat colostrum, and disease-free cow colostrum is a better choice than the dried products.

When you have a doe that kids, milk out some colostrum to heat-treat, freeze, and use in the future for an orphaned kid or one that needs it for another reason. Healthy older does can provide greater volume and quality of colostrum. A dam produces colostrum for up to four days after kidding, but the best colostrum to freeze for later is what she produces during the first day. After that, her milk begins to come in and she produces a combination of colostrum and milk.

Dealing with kid problems

Kid problems seem to come in spurts — you can go years without problems and then one year a series of kids develops the same problem. You can trace some of these problems, such as white muscle disease or diarrhea, to a specific cause, but frequently you find yourself baffled about the cause. All you can do is treat the symptoms and watch the other kids to see whether they are affected.

In this section, I tell you about some common — usually resolvable — problems that you may find in your newborn kids.

Chilled kid

On occasion, you'll miss a kidding or a kid will get separated from its mother after birth and get chilled. Because kids can't maintain their body temperature in cold weather and can't digest food when chilled, it's essential to get a chilled kid warmed up as soon as possible. To warm a chilled kid, follow these steps:

1. **Fill a sink with 100°F water.**
2. **Put the kid up to the neck in the water (you can put the kid in a large plastic bag first to avoid it getting wet or being stressed by blow-drying).**
3. **While keeping the kid's head above water, gently rub the kid's body and intermittently check for warmth by putting a finger in the kid's mouth.**
4. **When the kid seems warm, remove it and take a rectal temperature.**

   If the temperature is at least 100°F, towel-dry or dry with a blow dryer, being careful not to burn the kid.

   If the temperature isn't at least 100°F, put the kid back in the water.
5. **After the kid is warm enough, return it to the mother, with supervision, or bottle- or tube-feed it.**

Floppy kid syndrome

Floppy kid syndrome (FKS) is a disease in kids from 3 to 10 days old that causes sudden weakness and inability to move the legs. It was first documented in 1987, but its cause is still unknown.

FKS is more common later in a kidding season and can spread rapidly through the kids. The kids seem normal at birth and then suddenly don't want to nurse and have very weak muscles; they stumble like they're slightly drunk and eventually become limp as a wet dishrag. If you pick up one of these kids and gently shake her, you can hear liquid sloshing in her belly.

TIP

Treat any kid that you think might have this problem by giving a half-teaspoon of baking soda in 4 ounces of water. You can give this by bottle, but if kids refuse or are very weak, give it by tube-feeding. If the kid gets better within two hours, you can conclude that it had FKS.

Some kids get better and then relapse. Even if a kid responds to the baking soda, remove her from her mother, make sure she is warm, and keep her hydrated by giving electrolytes in a bottle for a day or two before resuming milk feeding.

Some kids develop secondary enterotoxemia, which must be treated with CDT antitoxin, or pneumonia, which requires antibiotics. (See Chapter 11 for more about these conditions.)

White muscle disease

White muscle disease is caused by a deficiency of selenium or vitamin E. (Chapter 11 tells you more about selenium deficiency.) Signs of the disease include weakness, stiffness, and rapid difficult breathing. The disease is most common in kids less than two months old. Because it can cause sudden death, white muscle disease can be mistaken for another disease, such as parasitism or enterotoxemia.

You can prevent white muscle disease by injecting the dam with BoSe prior to kidding and by giving an injection to the kid within a month after birth. As with FKS, you may only know that a kid has white muscle disease when he responds to the treatment. Otherwise you may not know until you have a necropsy done on a kid that has died.

WARNING

Talk to your vet or county extension agent to find out whether you live in an area that is deficient in selenium before routinely giving your goats BoSe because it can be toxic.

Navel ill

Navel ill (also called *joint ill* or *polyarthritis*) is an infection caused by bacteria entering the kid's bloodstream through the navel. Goats can get navel ill during their first month of life. Symptoms include lameness, inflamed painful joints, a swollen navel, and fever. The infection arises from unclean conditions and from not treating the umbilical cord with iodine right after birth (see the section "Caring for newborn kids" earlier in this chapter).

The infection can spread to the joints from the bloodstream, which is called *joint ill.* To treat joint ill, soak the navel area with a clean, hot, wet cloth, remove scabs or pus, and apply iodine. Contact your veterinarian about which antibiotic to use.

Hyperflexed legs

Every so often a kid is born with legs that seem to bend backward at the knee or that can't seem to lock the joints. This condition is called *hyperflexion.* It usually corrects itself within a day or two. In cases where a kid doesn't seem to be getting better, you can splint the legs by using a toilet paper roll or cardboard, duct tape or medical tape, and a little creativity.

Diarrhea (scours)

Newborn kids frequently get diarrhea (called *scours* in goats) because their immune systems are not yet developed. Kids that don't ingest colostrum right after birth are more likely to get scours, and so are kids that live in unsanitary conditions or have inadequate diets. Inadequate diets include underfeeding, overfeeding, and poor quality food.

When you find a kid with scours, evaluate whether he may have eaten too much grain or eaten something else that might be causing the problem. Eating too much grain or drinking too much milk can cause a kid to get *enterotoxemia,* which is an overgrowth of certain bacteria in the gut. (See Chapter 11 for more on enterotoxemia.)

At the first sign of diarrhea, give the kid kaolin pectin, collect feces for analysis, and watch the kid carefully for the next few hours. The fecal exam will tell whether the kid has *coccidiosis* or is overrun with parasites and needs to be treated. (Chapter 11 tells you more about these common ailments.) If the diarrhea continues or gets worse, contact your veterinarian. Because enterotoxemia can kill a kid quickly, many veterinarians recommend giving CDT antitoxin, even if the kid or his dam have received the CDT vaccine.

You not only want to fight the cause of scours but to make sure that the kid stays hydrated. A goat kid can dehydrate fast, leading to death. (Chapter 10 tells you how to check for and treat dehydration.)

TIP

Have the goat drink the electrolyte mixture or give it with a feeding tube. Give a large goat a pint and a kid or smaller goat ½ to 1 cup every six hours as needed. Also make sure that the kid has access to roughage such as willow branches, salal, blackberry, or whatever grows in your area.

Most of the time, the scours are transient and go away within a day. Kids just have sensitive tummies, and the slightest change can cause diarrhea.

Milk goiter

Milk goiter is a soft swelling of the thymus gland, which is located on the neck. It is often mistaken for an enlarged thyroid, which is caused by iodine deficiency.

Milk goiter is not a problem in itself, but it can alarm goat owners and even veterinarians who aren't familiar with it. Milk goiter is found in well-nourished kids, most commonly Nubians, but its function isn't known. The first kid born on my farm had it. She was a piggy little girl who nursed for a long time. She had the swelling but no symptoms other than being overweight for her first year.

If you have a nursing kid with a soft swelling on the front of her neck, with no other symptoms, think of milk goiter. There is no good way to distinguish milk goiter from enlarged thyroid, but if you think the kid might have iodine deficiency, paint the tail web with 7% iodine and see if it helps. The treatment for milk goiter is to do nothing and wait for it to subside as the kid gets older.

Tube-feeding a weak kid

When a kid is born too weak, is unable to suck, or is too sick to nurse or drink out of a bottle, you need to get fluids into him to keep him alive. If the kid is a newborn, he needs colostrum. To boost the kid's energy, add some corn syrup or Nutridrench to the colostrum.

For a weak kid that has already gotten colostrum, use electrolytes, B vitamins, probiotics, and goat milk or milk replacer. I keep on hand dried electrolytes I got from my vet. If the goat is being tube-fed because of scours, the electrolytes give him energy and give the gut time to heal.

TIP

Stay away from milk or milk replacer for at least a day and never mix milk and water or electrolytes, to prevent digestive problems.

You may have to tube feed a weak kid only one time to get it up and energized. If you need to tube-feed a kid more than once, do it only every 2 to 4 hours with the same small amount. Frequent, small feedings are better than infrequent large feedings.

To tube-feed a kid, you need the following equipment:

- Feeding tube
- 60 cc syringe with an irrigation tip
- Bowl of clean warm water
- 6 cc syringe
- A helper

To tube-feed a kid, take the following steps:

1. **Measure the distance you need to insert the tube so it ends up in the kid's stomach.**

   Measure from the nose to the center of the ear. Then measure from the ear down to the chest floor. Add the two measurements and mark the tube at that point.

2. **Have someone hold the kid securely so she is sitting up, not on her side.**
3. **Hold the kid's head straight up so the bottom of the chin and front of the neck are in a straight line (see Figure 13-4).**

   If necessary, fold up a towel to prop up the kid's head. This prevents aspiration, in the event the kid coughs up any fluids.

FIGURE 13-4: Keep a kid's head up when you tube-feed it.

4. **Dip the end of the tube in the warm water to soften it, and then insert the tube into the kid's mouth, over the tongue and down the throat until the length you marked is all the way in.**

   You may be able to feel the tube as it passes down the esophagus. Very weak kids won't even struggle; bigger ones may fight you.

WARNING

   If the kid was crying before you inserted the tube and suddenly stops during the process, pull it out until the kid can cry. Then try again. Tube feeding into the lungs has the same effect as syringing when the kid chokes — the kid can get pneumonia or die.

5. **Determine whether the tube is inserted correctly by using one of these methods:**
   - Smell the end of the tube for the milk smell of the stomach. This obviously won't work for newborns that haven't had milk yet.

- Listen at the end of the tube for little crackles. If you hear breath sounds, withdraw the tube and start over.
- Place the end of the tube into a cup of water. If it blows bubbles, you are in the lungs and need to try again.

6. **Inject about 5 cc of water into the tube with the 6 cc syringe to make sure it flows down the tube.**

   If not, withdraw it a few inches and try again. This step helps you make sure the tube isn't against the stomach wall or twisted or blocked in some way.

7. **After you have established that the tube is properly placed and the tube works, tube-feed the kid.**

   Put 2 to 4 ounces of the feeding liquid into the syringe and attach the syringe to the end of the tube. Don't use the inside part of the syringe; gravity will deliver the fluid to the stomach if you hold the syringe above the kid.

8. **After administering the fluids, rinse the tube by adding up to another 10 cc of water to the syringe.**

   This step is not critical, but it helps prevent the kid from aspirating the milk or electrolytes when the tube is being removed.

9. **Remove the syringe and tube.**

   Remove the syringe from the end of the tube. Then remove the tube slowly, with a finger over the end. Putting your finger over the end helps prevent excess fluids from getting into the lungs. Removing the tube too quickly can cause the kid some discomfort and possible tissue damage.

10. **Clean and sterilize your supplies after each feeding.**

Vaccinating

Most veterinarians recommend that you vaccinate kids with CDT (*clostridium perfringens* Type C & D and tetanus toxoid). This vaccine gives them some protection against enterotoxemia and tetanus. (See Chapter 10 for more on vaccinations.)

TIP

If you choose to vaccinate your kids with CDT and their dam was vaccinated in the month before kidding, you don't need to vaccinate the kids until they are six to eight weeks old. Give them a booster one month later.

If their dam was not vaccinated and you want to begin vaccinating, give the first vaccination when they are three weeks old and a booster a month later.

Feeding the Kids

TIP

Before your kids are born, figure out what and how you want to feed them. You have a lot of choices to make — bottle- or dam-raised, individual or group feeding, milk or milk replacer. Their little tummies don't like change, so make a plan and stick to it (unless you have good reason to change the plan, such as switching to bottle if the mom gets sick or switching to milk if a kid develops diarrhea on milk replacer).

Kids nurse for at least the first 8 weeks of their lives and gradually switch over to hay and grain. They start tasting hay and grain as early as the first week, but they need to be eating solid food well before weaning. This gives you a little extra time to decide on and get a supply of solid food.

Deciding between hand-feeding or natural feeding

Dam-raising goats is the easiest way to feed kids after you're sure that both parties have figured out how to do it. You just have to feed the mom and you don't have to manage a frequent feeding schedule. But things don't always go as simply as planned.

If a doe has more than three kids, you usually need to hand-feed a couple of them to make sure they are getting enough to grow and thrive. The doe may not have enough milk to go around or her health may suffer from feeding all those little mouths.

Another reason you may decide not to dam-raise is that a doe has a contagious disease that you don't want the kids to get. In this case, you will have pulled the babies at birth anyway.

Some people prefer to hand-feed their kids for other reasons: They want to prevent them from getting a disease the doe *might* have or they prefer the friendliness of hand-fed babies. You can raise kids that aren't hand-fed to be friendly, too — it just takes more time, but these goats are genuinely friendly, not just coming around because they think you have food.

TIP

If you have the time and are milking the dam anyway, put some of the milk in a bottle after the morning milking and feed it to a hungry kid that has been locked up all night without his mom. (He may be resistant at first, but after he figures out that milk is in the bottle, he will suck it down.) This intermittent bottle-feeding has the benefit of associating you with food, which will make him tamer and friendlier, but not annoyingly so.

If you decide to hand-feed, you need to decide what method to use and what to feed.

You may want to keep a newborn kid in the house for bottle-feeding, but kids should really be kept with other goats unless you cannot otherwise protect them or check on them frequently. When it's safe, move them to the barn. Kids need to learn how to be goats, and if they aren't raised around other goats, they may have a difficult adjustment when they finally do have to live with other goats. For example, they're more likely to be bullied and to cry a lot for the human who raised them.

Choosing milk

Raw goat milk is the ideal food for a goat kid. Raw milk helps with immunity against diseases, and it contains the good bacteria a new kid needs to populate his gut. If the dam is healthy, raw is best.

If the dam has CAEV or another transmissible disease, or if you are pooling the milk to be fed to kids, pasteurize it first. To pasteurize milk bring it to 165° Fahrenheit for 15 seconds. Cool as quickly as possible.

TIP

If you don't have access to goat milk, or if you want all of the milk for your own uses, you can feed your kids cow milk or milk replacer. When feeding cow milk to kids, add some half and half because cow milk doesn't have as much fat as goat milk. Feeding cow milk to kids can get expensive fast. You also need to make sure that the milk is free from Johne's disease — something you might not be able to do when you buy commercial milk. (See Chapter 11 for more about Johne's disease.)

If you use milk replacer, make sure that it is formulated for goat kids or, if a goat formulation isn't available, use lamb milk replacer. Lamb milk replacer needs to be reconstituted using a bit more water because ewe's milk is higher in fat. You can also mix milk replacer with goat milk.

WARNING

Although some people use milk replacer successfully, I haven't had good luck with it and don't recommend using it. More of my kids got diarrhea with it than with their mothers' milk. I also prefer the ease of pouring cow milk rather than having to mix it up.

Feeding individually or as a group

If you have only a few kids to feed, you can feed each one with an empty soda bottle with a *Pritchard nipple* (a red, soft latex nipple designed to replicate a ewe's teat) screwed on the top. To get a kid to drink from a bottle, take the following steps:

1. **Hold the kid on your lap between your legs with its head up in the position he would use when nursing.**
2. **Hold the bottle in your hand and slide it into the kid's mouth.**

   You may have to pry the kid's mouth open the first few times.
3. **Wrap your hand around the bottom of the kid's muzzle, making sure to leave his nose clear, and hold his mouth closed while also holding the bottle in place.**

   After the kid gets the hang of nursing from the bottle, you can just hold it at an angle above the kid and he will take the nipple and nurse.

If you have a lot of kids, you may soon tire of the one-on-one feeding (and the other kids jumping all over you at the same time) and consider a way to feed more than one at a time. For larger groups of kids, you can buy a milk feeder that is basically a bucket with holes drilled in the side and nipples sticking out. You can get one of these bucket feeders at a feed store or from a goat supply catalog or, if you're handy, you can make one of your own.

Feeding a group of kids a mixture of milk from more than one doe can create potential risks for kids if any of the does contributing to that pool develops a health problem. Unless the milk is pasteurized, that doe has just potentially infected every kid. Unless you know that all of your does are healthy, pasteurize milk before pooling it.

You can also try feeding kids from a shallow pan. To get the kid started, dip her mouth gently into the milk. She will start to drink it after she tastes it. This is no different than drinking water out of a bucket, which kids pick up very easily.

Whatever method you choose, make sure to rinse the buckets, bottles, nipples, or other equipment with cool water and then wash and sanitize them after each use to prevent the kids from getting sick.

Feeding schedule

Different breeds (and sizes) of goats eat different amounts of food or milk, so in Table 13-1, I give you general guidelines for feeding milk or milk replacer. This schedule applies only to kids that are hand-fed because those that are dam-raised set their own schedules. Give miniature breeds about half the amount recommended here.

As you decrease the frequency and amount of milk, the kids gradually shift to hay and grain.

TABLE 13-1 A Sample Goat Feeding Schedule

Time Frame	Amount	Frequency
First two days	4 ounces	Every 6 hours
Days 3 through 7	8 ounces	Every 6 hours
Second through fourth weeks	10 ounces	Every 8 hours
Fifth through eighth weeks	12 ounces	Every 12 hours
Ninth through tenth weeks	6 ounces	Morning and night

Introducing solid foods

Kids learn to eat by mimicking their mothers. Those who live with their dams often start nibbling on hay or straw at only a few days of age. Kids that are raised separately might take a little longer to figure out that hay is food.

Regardless of where they live, make sure the kids have hay at all times so they can eat it whenever they want. Hay helps their rumens develop.

If you feed grain to the whole herd at the same time, the kids will be intrigued and jump right in to find out what tastes so good. Some will catch on more quickly than others — these are the ones to watch for diarrhea and enterotoxemia due to overconsumption. (See Chapter 11 for more about feed-related diseases.)

To prevent other goats from eating the kids' grain, you need to create a *creep feeder* — an area that the little ones can get into but the big ones can't. You can make a separate area with cattle panels, which kids can squeeze through, or you can put a cattle panel over the front of a stall. Keep some grain in a bowl and try some other treats like sunflower seeds or kelp. Start very gradually with only a small handful of grain per kid each day, with a maximum of a pound (2 cups).

Check the kids regularly; they're getting too fat if you can't feel their ribs. If that's the case, you need to decrease their grain. Make sure the kids always have access to clean, fresh water, too.

Weaning kids

The thought of weaning kids makes me laugh because the mom and kids may have a different idea. If you don't have an extra space to move the kids to, you may just have to give up your dream of weaning at a certain time. You can separate mom and kids at night so that you can milk in the morning, but during the day the mother-child pair can do what it wants. They usually cut down on the nursing substantially, but I still see them sneaking a drink from time to time.

In a case early on in my goat-raising days, I had a kid who tried to nurse her mom when the poor doe was in labor. (Someone had told me that the kidding doe would feel better to have a friend in the kidding pen, something I have generally found not to be the case.)

So, how do you wean a kid? If the kid is on a bottle, weaning is easy. Just quit feeding the milk. You will have to listen to some yelling from the kid for a few days until she realizes that crying won't work. I usually stop bottle feeding gradually, cutting down to once a day for a week and then stopping altogether.

In dam-raised kids, if you're lucky the mom will decide that she's finished and walk away when the kid tries to nurse. If you have a certain date in mind, though, you have to take things into your own hands and physically separate the kid. If you aren't milking the doe, then a month is usually adequate — her milk dries up, and when you put the kid back in nothing is there.

For a milker, the process can be a little more complicated. I once tried to wean a kid by moving her to another pen with some other kids that were also being weaned. They lived there for two months and then, thinking the habit was broken, I moved them back in with the does. Well, that little goat immediately rushed up to her mother and began nursing!

THEY MAY BE CUTE, BUT YOU CAN'T KEEP THEM ALL

So you got your first two does, had them bred, and you had two of the cutest sets of goat kids ever born. Now the goat area that has worked well for two has become home to another four goats that are growing before your very eyes. And you just love watching those little kids bounce around. So how are you going to deprive yourself?

If you're raising goats for meat the answer is easy: Raise them and then sell them or butcher them yourself. If they're fiber goats or dairy goats you may have the same solution, depending on your goals and your intestinal fortitude, but for a lot of people, selling those cute kids is hard.

Set a limit in advance of how many and what sex of kids you will keep. Then figure out where you can sell them and develop a plan for marketing them. (Chapter 15 gives you marketing ideas.)

Weaning is not always that complicated. Because I milk my goats, I start by locking nursing kids in a separate pen every night when they are about two weeks old. Then I get the morning milk. They have some grain, hay, and water in the pen, which helps them learn that they can get nourishment in other ways.

I like to leave things to the dam-kid pair to work out how long they will nurse. As a result, I sometimes have a fat kid, but usually they discontinue nursing no later than six months.

» Making life easier for your aging goats

» Knowing when it's time to let go

Chapter 14
Caring for Aged Goats

More and more people who are raising goats see them less as "just livestock" and more as friends or pets. Along with this philosophy often comes a commitment to caring for these creatures when they get old and beyond their productive years.

Although goats may live many more years, after the age of about 9 or 10 (and sometimes earlier) you can expect your best milker or favorite wether to start to show signs of aging. I've had does live as long as 14 years, and I have a friend whose wether lived to the ripe old age of 17. Neither were without problems, but both had good lives until the day they died. They were even a little bit spoiled — and deserved to be!

In this chapter, I cover the kinds of problems you may encounter in your aged goats, as well as ideas for making their lives as pleasant as possible until it's time to let go.

Identifying Health Problems

Geriatric goats can develop chronic health problems that aren't seen in younger goats. Over time, they may become more disabled and require more care until they die or have to be put down. In the following sections, I walk you through the various problems goats experience as they age. Your goats may not have all these problems but will like have some of them as they age.

One note about goats who have tested positive for Caprine Arthritis Encephalitis Virus (CAEV): A minority of goats with CAEV will maintain good health during their lives, but certain conditions may develop as they age — especially if they've previously shown symptoms. These issues include arthritis, chronic mastitis, weight loss, and problems related to the immune system. For more on CAEV, turn to Chapter 11.

Musculoskeletal issues

Older goats have a harder time keeping weight on and maintaining their body condition. They also tend to have more problems with their legs and feet, mainly because they're usually on their feet and their joints simply wear out, causing them to become lame and stiff or to develop arthritis.

Watch how your old goat walks. Is she limping? Is she moving more slowly? Does she seem uncomfortable when lying down or getting up? Does she look like she's in pain? These are all indicators that she may have arthritis.

You may also find that your goat's hooves are growing crooked and are harder to trim. As your goat ages, it's even more important to keep up with your routine hoof care. Trim the hooves of a geriatric goat at least monthly to help prevent deformities and keep them in good shape.

Standing on a milk stand and having a leg held up for trimming may be uncomfortable for her, so trimming frequently will help to minimize the time you have to do so.

Digestive issues

The digestive system (covered in Chapter 2) is key to maintaining body condition and staying warm. That's because the activity of the goat's stomach is what generates the energy needed to maintain the proper body temperature and a healthy weight.

An effective digestive system begins with the mouth and the teeth. Goat teeth can get worn out, break, or even fall out over time. If you notice an aged goat having trouble chewing, foaming at the mouth when trying to eat grain, "cheeking" hay, dropping a wad of cud, or just losing condition, think about the possibility of dental problems.

Sometimes you can identify tooth loss by feeling the goat's face. If you want to be sure, have your vet check for tooth problems to determine whether the teeth need to be filed down (called *floating*) or even pulled.

TIP

If your goat seems to not to be interested in food, you can give vitamin B complex to stimulate the appetite.

Older goats are also more susceptible to internal parasites, so consider doing fecal exams more frequently. (See Chapter 11 for more on internal parasites.)

Immune system issues

One part of aging is a decrease in the effectiveness of the immune system. What this means for your aged goat is a lessened ability to fight off infection, parasites, and respiratory or circulatory problems.

Old goats also have a hard time maintaining a healthy coat and are more prone to have a rougher coat, hair loss, and chronic skin disease. They're also more prone to external parasites, such as lice and mites, so check for these issues first when evaluating the coat and skin. (I tell you about identifying and controlling external parasites in Chapter 11.)

If you don't find any lice or mites, you can have a vet check for thyroid function or suggest a strategy for improving the skin and coat. I have found that feeding kelp meal and black oil sunflower seeds, which add vitamins and minerals to the diet, can help to maintain a healthy coat.

Circulatory issues

A goat's organs, particularly the heart and lungs, can begin to develop problems as the goat ages. This can affect a goat's willingness to get up and be as energetic as he once was. Goats may develop a chronic cough or have problems breathing, too. Because of the decreased effectiveness of the immune system, they can also get pneumonia.

TIP

If you have a goat who is weakening this way, remember that she also can't handle as much stress and will be affected more severely when bullied by younger goats, so providing her with a separate living area is even more important (see "Providing proper housing," later in this chapter, for more information).

Mammary issues

A goat with CAEV (see Chapter 11) is more likely to develop chronic *mastitis* (inflammation of the udder) than one who started out healthy. But even healthy goats can develop mastitis. (Chapter 15 tells you about the signs of mastitis and how to test for and prevent it.)

Chronic mastitis can lead to a hardening of the udder and even gangrene in some cases. You need to be alert to this possibility, especially if your older doe was a high producer or had repeated bouts with mastitis. Options for treating it include local treatment with antibiotics or even a mastectomy, if you can afford it and your vet agrees.

Behavioral changes

As a goat ages, you'll inevitably see behavioral changes. Herd queens often lose their status in the herd as they become weaker and less able to defend their position. I've usually seen this play out over a period of time. As a herd queen gets challenged by younger does who aspire to move up in the herd, she may get tired of fighting and gradually give up her leadership position.

Depression is another change that occurs in aging goats. It can be part of a disease process — I've seen it with pneumonia in goats — or it can be due to a loss. This is one reason that isolating an older goat can cause more problems if you don't consider who their companions are and how to meet those needs (see "Providing proper housing," later in this chapter, for more information).

Making Allowances for Disability

Keeping your old goat comfortable is key to ensuring that what is left of her life is worthwhile. Goats can't tell you what they're feeling, so it's up to you to try to read their symptoms and do what you can to prevent problems before they start.

Providing proper housing

Proper housing is essential for an elderly goat, so you need to think about whether your current setup will work or if you need to make some modifications.

TIP

I have two kidding/infirmary stalls that I use for goats who are impaired. If you do need to make changes to your current setup, here are some of things to consider when doing so:

- **Older goats may be unable to maintain a normal body temperature.** If you find that your geriatric goat is cold (with a low body temperature) or shivering or gets too hot in the summer, create or use an existing smaller stall to house her. This area will need plenty of dry bedding and must be protected from drafts.

In colder weather, you can provide a goat coat (see Chapter 2) or a neck gaiter to help keep her warm. I used a polar fleece gaiter from a friend for my 14-year-old goat Celtic Kid during the winter.

When using a coat on an elderly goat, make sure to check it regularly to ensure that it hasn't gotten wet with urine or that external parasites haven't taken up residence.

If you can locate a stall where electricity is available, you can use it for a swine farrowing mat that can be purchased from a livestock supply store or a flat plastic electric warmer from a pet-supply catalog. These are safe to put under the hay and work well to keep an old doe warm.

In the summer, shaving the neck, udder, and stomach will help a goat stay cooler. A fan is also helpful, if you have electricity available.

- **Friends are important.** When planning housing, think about that special friend or daughter who your old goat is used to hanging out with and sleeping with and plan the size accordingly. They may not need to spend *all* their time together, especially if the younger one has a lot of energy and wants to get out with the more rambunctious goats during the day, but having a friend can help combat depression.

  Cattle or hog panels are also a good tool for separating goats for their safety. They have the advantage of allowing a goat to be safe while right next to her friends, because they aren't a solid wall and don't feel so isolating. Also, if your elderly doe is one who is particularly fond of kids, a cattle or hog panel will allow kids to get in but keep bigger goats out.

- **Animals don't like weakness.** Another reason for special housing for the geriatric goat has to do with the nature of animals. Some of them are bullies. This is especially true when one of them is sick or weak. You've probably heard of a "pecking party." Well, it isn't just chickens who recognize weakness and act on it. Goats have a well-established hierarchy, and when they see an opportunity to increase their status, they usually act on it and bully the weaker one. Plan your housing to keep the bullies out.

Finding a workable diet

Finding a workable diet for an older goat requires determining not only *what* to feed but also *how* to feed it. If a geriatric goat is having trouble keeping weight on, and you've confirmed she's free of dental problems, doesn't have a parasite overload, and has no other health issues, you may need to may need to change where you feed her, adjust her diet, or both.

Changing the feeding location

You can use a separate stall or a milking room to feed an older goat away from the rest of the herd. It will take a little more time and effort than just putting out hay or grain and walking away, but it's worthwhile to keep her comfortable. Take your time when feeding her; she may need more time than a young goat does.

A separate feeding location can be important because younger, stronger goats may prevent an older, more frail goat from getting to the food. It also will allow you to tailor your older goat's diet to her different needs.

Some older goats with arthritis have problems with eating out of a feeder at normal height due to difficulty standing long enough or pain. You can feed these goats in a bowl or on the ground. However, remember that goats who eat on the ground are more likely to get parasites, so make sure it's in a clean area.

Adjusting the diet

Just feeding an older goat separately from the rest of the herd may resolve the problem of losing condition, because it ensures that she's getting her fair share. If this strategy doesn't improve her body condition, or if you've already identified a health problem that is causing the weight loss, you'll need to adjust her diet to ensure that she gets the nutrients she needs.

To help an elderly doe who is losing weight, you can add some chopped alfalfa to her diet. This will add calcium and protein and may help her get some condition back.

WARNING

Don't add calcium to the diet of an elderly wether. It may cause urinary calculi.

I've had good luck with adding senior horse feed to the ration of older goats. This should be given at a rate of 1 pound of feed for every 100 pounds of weight. If chewing is a problem, hydrate the feed with warm water first. Hydrated beet pulp or rice bran can also help maintain weight and the rumen in a goat that has trouble chewing.

Chaffhaye is another helpful substitute or addition to hay in the diet (see Chapter 6). Goats who have problems chewing often have an easier time with Chaffhaye than they do with hay, because it isn't so dry.

Goats appreciate having fresh *browse* (vegetation such as plant shoots, shrubs, or twigs) cut for and fed to them. My go-to for browse is salal and blackberry leaves. To help with putting on weight (or just make your goat happy), you can also give treats such as cut-up apples, pears, or other fruits; peanuts in the shell; or corn chips. Just don't overdo it.

In addition, if you see that an elderly goat doesn't appear to be chewing her cud, giving a probiotic, such as Probios, can help. If that doesn't work, you can steal a cud from another goat.

To steal a cud, take a clean bowl and a drenching syringe of warm water. When you see a healthy goat start chewing her cud, syringe the water through the side of her mouth into the back and catch the runoff in the bowl. This can then be pulled into the syringe and given to a goat that needs a cud transplant.

Water is critical to keeping a goat alive. If a geriatric goat loses interest in drinking and starts to get dehydrated, you can try flavoring the water or giving electrolytes. For a goat who doesn't want to drink, you may have to drench her (give her liquid with a drenching syringe, which you should have in your first-aid kit).

TIP

Don't forget that older goats still need access to minerals and other supplements that you normally feed. You can provide them with a separate mineral feeder inside the stall or feeding area.

Easing chronic pain

Chronic pain in older goats is usually related to arthritis. It can be severe if they have CAEV or an old injury that is starting to bother them, or it can be just garden-variety arthritis caused by worn-out joints.

In addition to the obvious signs of arthritis — such as difficulty getting up and down and moving around — grinding teeth is an indicator of pain.

If a specific joint is involved, talk to your vet about getting a specialized splint or using Vetrap on it. If you have the means, some specialized vets even offer or refer to professionals who provide chiropractic, acupuncture, and massage therapies. If you're interested in these approaches, talk to your vet about getting a referral.

TIP

You can work with your vet to treat pain with an over-the-counter nonsteroidal anti-inflammatory (NSAID) such as ibuprofen or a prescription drug like meloxicam. Also effective are some of the remedies used by humans with these problems, such as methylsulfonylmethane (MSM), glucosamine, and chondroitin.

Some people who prefer natural remedies have had success using turmeric or other herbal remedies.

Determining When to Let Go

Letting go of an animal is so difficult, but watching him suffer isn't easy either, nor is it fair. In an emergency situation, making the decision to let a goat go is often easier than it is when a goat is gradually failing.

For the sake of your beloved goat, you need to watch her closely. If you pay attention, you'll know when it's time to make the call. The hard truth is that this decision may be financial — or at least partly so — especially when the only solution to extending life leads to a cost-prohibitive vet bill.

There are a number of signs that your goat will be better off if her suffering is ended. They all relate to quality of life. If they persist, the goat's life will become less meaningful and tolerable over time. Watch closely for the following:

- **Uncontrolled pain:** Watch for signs that the goat is grinding her teeth and still appears to have uncontrolled pain, despite any drugs being given to treat it.
- **Loss of appetite:** If you've adjusted her diet and tried all the suggestions for stimulating appetite, but your goat still doesn't want to eat, she won't last long.
- **Lack of interest in drinking:** Having an older goat lose interest in drinking, despite giving her electrolytes or even an intravenous (IV) infusion of fluids, that's a sure sign that she's ready to go.
- **Immobility:** When a goat is unable to stand up, he'll gradually lose the ability to ever do so because muscles atrophy. This can also lead to other problems, such as soiling due to urine and feces. Consider the feasibility of using a sling or helpers to get him up and about.
- **General unhappiness:** You know your goat and can tell when she's happy to see you. She'll make eye contact, eat treats, and respond to touch. You can also observe how she interacts with friends or whether she's withdrawing from them as well. If she's depressed and disinterested in life, you need to consider whether it's fair to ask her to continue.

Goats can't plan for tomorrow; all they know is what's happening right now. When your beloved goat is suffering from many or all of these signs over a period of time, you need to seriously consider whether waiting for her to die naturally or having her euthanized is more humane. This is one of the hard realities of owning any animal. We're entrusted with the lives of these creatures, and we have to make the decisions that are right for them and for us.

If you're really struggling with the decision, talk with your veterinarian or another goat-owning friend. Sometimes getting some input from other folks can help you feel confident in your decision.

4 Making Your Goats Work for You

IN THIS PART . . .

Milk your goats and sell their milk.

Get to know the meat goat market and find out where to market your goats.

Show your goats and rack up the awards.

Make yarn, go backpacking, board other people's goats, and more.

» Troubleshooting milking problems

» Using and selling goats' milk

Chapter 15
Discovering Goat Milk: How to Get, Use, and Sell It

If you've ever tasted fresh, cold goat milk, you know that the white stuff in stores can't compare! You can drink goat milk raw or pasteurized, or you can make it into an endless variety of cheeses, yogurt, and other dairy products.

Goat milk is high in vitamin A, butterfat, and protein. It is more digestible than cow milk because the fat globules are smaller. Just about anyone you talk to knows of someone (usually a baby) who was "allergic" to cow milk and whose problem resolved when they switched to goat milk.

Getting milk from goat to table takes time and care. If you're a novice milker, expect a learning curve before you and your animals are completely comfortable with the routine. And if you expect to supplement your income (and the goats' feed) by selling milk, you might be in for a rude awakening depending on the laws of the state where you live.

In this chapter, I give you an overview of the laws regarding selling milk in the United States. I also discuss hand- and machine-milking, how to handle the milk, and how to make various products from the milk.

Developing a Milking Routine

Goats are creatures of habit. If you want to maximize the amount of milk you get and make milking easier, you need to develop a regular milking routine, which means using the same place and same procedure every day.

A milking routine requires you to

- Have a milking area separate from the other goats (otherwise they will bug you and steal grain from the goat being milked).
- Milk the goat from the same side every time.
- Wash the udder first, to encourage the udder to *let down* (release the milk) and to ensure cleanliness.
- Milk your goats in the same order each time, unless one gets mastitis (an inflammation of the udder, often caused by bacteria; see the upcoming section "Preventing mastitis"). You can choose any order you want, but I have found that the goats usually choose the order, with the *herd queen* (most dominant) going first. If you have CAEV-positive goats, milk them last or use separate equipment (see Chapter 11).

Knowing when to milk

Other than the rare *precocious milker* (a doe that has udder development and milk production without kidding), in order to *freshen* (produce milk), a goat first has to have a kid. This doesn't mean that she has to have a kid every year, though. Some people successfully milk their goats for years without rebreeding. I have done this with my goats for three or four years at a stretch.

Before she produces true milk, a goat produces *colostrum*, a rich, immune-system-boosting fluid that kids need during their first days after birth. (See Chapter 13 to find out more about colostrum.) The supply of milk that a goat produces is based on the demand for that milk. If you milk only once a day, you get less milk than you do with two milkings because the goat produces less (unless she has kids still nursing and creating demand). You can milk three times a day and get even more milk, but doing so is generally not cost-effective when you consider the amount of time it takes.

Unless you're bottle-feeding kids, let them nurse whenever they want for the first two weeks. Then put the kids in a separate area each night and milk their *dams* (mothers) in the morning before letting them out for the day. The kids keep the dams milked during the day, although you can usually get a little bit from an evening milking. That is, until they learn the routine and rush to their mothers to get that last drop before being locked up.

If you are bottle-feeding and plan to use the doe's milk you need to start milking right after they kid. You can take their colostrum for feeding the kids or freeze for later use.

TIP

To prepare the does for twice-daily milking, put them through the milking routine in the evening whether you milk them or not. That way you can gradually increase their tolerance to grain in anticipation of weaning the kids and milking twice a day. (To understand why you need to gradually increase their tolerance to grain, see Chapter 11 on enterotoxemia and bloat.) Whether you milk once or twice a day, you need to do it at the same time to ensure that does don't decrease production or get uncomfortable from a too-full udder.

Keeping the milk fresh

The axiom "You can't make a silk purse out of a sow's ear" applies to goat milk as well: You can't make fresh cheese out of sour milk. The best way to keep milk fresh is to drink it as soon as possible. But even before that, whether you pasteurize your milk or drink it raw, you need to start out with the best quality possible.

Starting with good milk

Starting with good milk means keeping it free of bad bacteria and other contaminants when you milk. To do this, you need to

- Make sure that the goat and her udder are clean.
- Milk in a clean environment; make sure your hands and equipment are clean.
- Strain the milk into a glass or stainless steel storage container right after milking.
- Chill the milk as soon as possible after milking to inhibit the growth of bacteria in the milk that contribute to spoilage. (See the upcoming section "Cooling the milk" to find out more about chilling milk.)
- Keep the milk out of direct sunlight or fluorescent lights, which can lead to off-flavors and loss of nutrients.

Pasteurizing

Pasteurization is the heating of milk to destroy bacteria and other harmful organisms. It extends milk's shelf life and is also the only way that you can make milk from a goat with a chronic disease safe for feeding to kids.

Unfortunately, pasteurization also destroys good organisms rather than just targeting the bad ones. And it changes the flavor of the milk, and of cheese made from pasteurized milk.

Like a lot of people, I prefer my goat milk raw because I know that my goats are healthy, I don't use antibiotics or other drugs on my milkers, and I handle the milk carefully. At times I have pasteurized milk because it was a from a goat with CAEV but I wanted to feed it to bottle babies. (See Chapter 11 for more information on CAEV.)

To pasteurize milk, follow these steps:

1. **Put milk in a double boiler or in jars in a pasteurizer or canner and heat to 165°F for 15 seconds.**
2. **Cool the milk as quickly as possible, but make sure not to put the hot jars in cold water or they will break.**
3. **Store the milk in the refrigerator.**

Caring for the udder

The udder is made of delicate tissue (although it doesn't seem delicate when you see goat kids bashing it to let down the milk). The udder is comprised of two halves and is held up by ligaments in the front, back, and sides. Each half has a mammary gland and one teat. Most of the milk is stored in the mammary gland until the udder is stimulated to let it down for kids or for you at milking time.

Good udder care includes these practices:

- Following a routine and properly milking the goat to avoid overfilling of or injury to the udder.
- Washing and drying the udder and teats before milking to minimize bacteria.
- Sanitizing the teats to prevent bacteria from entering the teat canal after milking.
- Making food available right after milking to encourage the goat to stand for a while after milking and allow teat canals to close.
- Promptly caring for an udder injury if it occurs. Wash a cut or scrape with warm, soapy water and keep an eye on any injury for complications such as mastitis.

Preventing mastitis

Mastitis is an inflammation of the mammary gland. It is usually caused by bacteria, but also may be the result of CAEV. (See Chapter 11 for more on CAEV.) Mastitis is more common in older goats that have developed saggy udders, which are more easily injured and exposed to bacteria.

Keep an eye out for the signs of mastitis, which include hot, swollen udder; fever; loss of appetite and energy; bloody, stringy, or bad-smelling or -tasting milk; and hard udder. Depending on the severity, the doe may have no signs at all.

You can identify and treat mastitis before it becomes severe by routinely testing with one of the following two tests, which are available through goat or cow supply catalogs. These simple and inexpensive tests identify white cells that signal infection:

- **California Mastitis Test (CMT):** To use the CMT, just milk a few squirts from each side into a different section of a plastic paddle and add the CMT solution. Then swirl it around and, if infection is present, the texture changes.
- **PortaSCC:** The PortaSCC subclinical mastitis test consists of packets with two test strips, pipettes, a color chart, and a bottle of solution. You simply put a drop of milk from each teat on each of the test strips and add three drops of the solution. Then wait 45 minutes and compare the color to the chart to identify problems. There is also a 5-minute version, but it isn't recommended with high-fat milk, which includes goat milk.

You can help prevent mastitis by properly milking and caring for the udder, regularly cleaning areas where goats spend time lying down, and not bringing goats with contagious diseases into the herd.

Keeping records

Tracking milk production (normally by weight) is a good way to figure out how much milk your goats are producing from year to year and whether their production is increasing, decreasing, or staying the same. It also can provide a clue that a goat is sick, because her production goes down.

Milking records can range from informally recording milk weights on a piece of paper to formally enrolling in a Dairy Herd Improvement Association (DHIA) program and receiving monthly reports and records for each doe.

Each month participants in DHIA programs send in a sample of milk, along with the weight of the milk from each goat over a 24-hour period, and receive monthly reports that provide information on the milk, such as butterfat and protein percentages and somatic cell counts (a key to whether the goat has mastitis).

Ending milking (Drying off)

When you are tired of milking, have to go on vacation, or a doe is three months into her pregnancy, you need to dry her off — that is, stop milking her.

There are two ways to dry off a doe: Cut back from milking twice a day to once a day and then stop milking altogether, or just stop. She may be ready to stop, too — especially if it is winter and she has been milking for a long time. At first her udder may swell as she accumulates milk that is not going anywhere, so you might be tempted to milk out a little at a time. Don't do it. Doing nothing protects her from mastitis and is the best thing to do.

If you are drying off a doe because she has mastitis, buy an intramammary antibiotic, such as Tomorrow, and administer it into the affected udder half the last day you milk. Make sure not to start milking before the withholding time is up.

Some treatments are for drying off and some are for does in milk, and you need to be careful about which one you get. If the doe has an active case of mastitis, you're better off treating the mastitis *before* drying the doe off. If you dry her off while she still has active mastitis she is likely to *freshen* (start producing milk the next time she has kids) with mastitis again.

One thing to consider when deciding to purchase milking does is that ideally you need to milk them for at least ten months out of the year. Milking ability is a learned trait as well as a genetic one, so if you continually dry your does off early, they will begin to dry themselves off even if you decide you want to milk them longer this time. Early drying off will decrease the amount of milk the doe produces every milking and over her lifetime.

Getting a Grip on Hand-Milking

Hand-milking a goat isn't really difficult, but you do have to practice to be efficient and fast at it. Some goats are like cows and have teats that are large enough so you can use all fingers on them, while others are so small that you can only use three fingers.

Never pull on the teat. This is not how milk is extracted, and it can cause injury to the mammary system.

You milk a goat by using one of two methods. With either method, use your thumb and forefinger to form a ring around the top of the teat (which stops the milk from going back into the udder). In the all-fingers method, use *all* of your fingers to *gently* press the milk out of the teat. In the other method, gently squeeze with your second and third finger (or however many fingers will fit). Figure 15-1 shows you how to use your thumb and fingers to milk a goat.

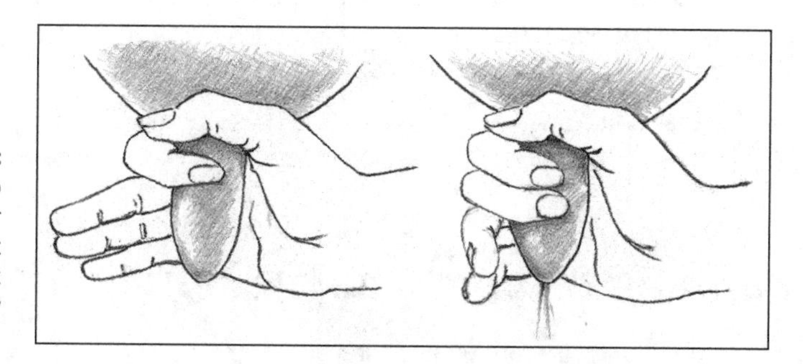

FIGURE 15-1: Wrap your thumb and forefinger around the teat to trap the milk and then gently squeeze it out.

TIP

To practice milking without fear of injuring the goat, use a rubber glove filled with water and tied shut at the top. This will give you an idea of how closing the teat (finger) off from the udder (hand) traps the water in the finger, allowing you to "milk" the liquid out by squeezing the teat.

Getting the supplies

You need few supplies to milk a goat. They include the following:

- **Milk stand:** Although people milk their goats in every situation imaginable — tied to the fence, on top of the dryer, in the kitchen — I recommend that you invest in a milk stand. A *milk stand* consists of a platform for the goat to stand on and a stanchion that latches and holds the goat's head. It gets the goats off the ground, restrains them, and gives them a familiar place to be milked. You can learn how to build your own milk stand in Chapter 4.
- **Stainless steel bucket:** Start with a six-quart bucket unless you are milking Nigerian Dwarves or Pygmies, which require a smaller one because they're shorter. You can also purchase an inexpensive stainless-steel bowl at a yard sale or thrift store for an uncooperative milker, and milk a little at a time and dump into the bucket frequently.
- **Udder-washing supplies:** I use an old plastic coffee can with hot water and dish soap (you can alternatively use a dairy-specific sanitizer), rags made from towels cut into smaller pieces, and paper towels for drying. You could also use dry towels or rags for drying. You need to wash the container after every milking and rinse with boiling water or a bleach solution (one part bleach to ten parts water). Use the towels only once and then rewash. As an alternative, some people use baby wipes.

- **Teat sanitizing supplies:** You need teat dip and cups (plastic containers that hold teat dip solution and have an opening that fits over the teat) or spray teat sanitizer, which you can purchase from a dairy supply company or feed store.
- **Stainless steel strainer and milk filters:** *Milk filters* are flat paper filters (somewhat like coffee filters) that fit into a strainer. You can buy strainers and filters from a dairy supply company or feed store.
- **Jars for milk storage:** Half-gallon mason jars with plastic lids work great, because the plastic doesn't rust when it gets wet, like metal does.

Running through the hand-milking process

To hand-milk a goat, follow these steps:

1. **Get the goat onto the milk stand and secure her in the stanchion with some grain for her to eat.**
2. **Wash your hands.**
3. **Clean the udder and teats with warm water and soap or sanitize with a wipe such as Milk Check Teat Wipes and dry them with a clean paper towel.**

   Make sure to thoroughly dry your hands.
4. **Wrap your fingers and thumb around each teat to trap some milk in the teat (see Figure 15-1) and squeeze to quickly milk one or two squirts from each teat into a cup.**

   This step allows you to check for abnormalities and removes any milk close to the surface of the teat that is more likely to be contaminated with bacteria. If the milk is abnormal, dispose of it after milking. (See the section "Caring for the udder".)
5. **Promptly milk the goat into a sanitized bucket, being careful not to pull on the teats.**

   The less time you spend between washing and milking, the higher the milk yield. If you take too long to milk, don't be surprised if the goat starts dancing or causing other mischief if she runs out of grain. You will have to give her seconds, making an effort not to overfeed her until she is used to larger portions.
6. **When you think the udder is empty, massage the back and bottom of the udder and bump it gently with your fist (more gently than the babies do!) in the front near the teats to encourage further let-down.**

7. **Pour the milk through a clean, filtered strainer into a clean jar.**
8. **Dip or spray the teats with a sanitizer such as Derma Sept Teat Dip of Fight Bac.**

   If you use dip cups, use a clean one for each goat to avoid cross-contamination.
9. **Return the goat to the herd.**

   Make sure you have some fresh hay or alfalfa and fresh water available for the goat right after milking. She will eat and drink instead of lying down and exposing an open teat orifice to bacteria. The orifice gradually seals, protecting the goat from mastitis.
10. **Clean the bucket and strainer.**

    Rinse the bucket and strainer with tepid water right away. Wash with warm soapy water and rinse with boiling water or a solution of one part bleach to ten parts water and air dry. If you have a dishwasher, wash the buckets and jars on the sanitize cycle. To remove milk stone buildup (residue caused by minerals in milk), use a solution of half white vinegar and half warm water.

You are likely to encounter problems with milking from time to time, especially if you are new to milking. Be aware that the first few days after freshening, your goat's milk will still have colostrum in it, which can affect the quality. Table 15-1 describes some other problems that you may encounter and how to solve them.

TABLE 15-1 Solving Common Milking Problems

Problem	Solution
Blood in milk	Do a mastitis test (see the section "Preventing mastitis," earlier in this chapter). If first freshener, it may be a broken blood vessel. Continue milking and discard milk. Wait to see whether is disappears within a few days.
Problem with let-down	Let kid nurse first. (More common in first fresheners and goats that are nursing kids.)
Off-flavor milk	Do a mastitis test. If negative consider illness, odor (bucks), or recent food consumed (onions, garlic).

Using a hand-milking machine

TIP

Goat milkers who have developed pain and other problems with their hands may want to try a hand-milking machine such as the Udderly EZ Milker (`www.udderlyezllc.com`). Originally invented to extract colostrum from mares after foaling, these hand-pump machines have been modified for milking goats.

The hand-milking machines consist of a simple hand vacuum pump, a silicon insert, tubing, and a collection jar. You attach the insert to one teat at a time, pump to get a vacuum, and then milk then goes down the tube to a collection jar. The machines eliminate the need for straining the milk and are said to be much easier on your hands.

To clean a hand-milking machine, just rinse the tubing with tepid water right away. Wash the machine and tubing with warm soapy water and rinse with boiling water or a solution of one part bleach to ten parts water and air dry. To remove milk stone buildup (residue caused by minerals in milk), soak in a solution of half white vinegar and half warm water.

WARNING

Using a hand-milking machine without the pulsation can be hard on a goat's teats. Some experienced dairy goat owners recommend only using it as backup and not regularly.

Using a Motorized Milking Machine

If you don't think your hands are strong enough to hand milk, you have too many goats, you have a problem such as carpal tunnel, or milking your goats by hand just takes too long, consider getting a milking machine. Figure 15-2 shows you a pulsating milk machine, which is the type that small-scale dairy goat farms use. Rather than one with a direct line into a bulk milk tank, it milks directly into jars.

A milking machine is a combination of several components that work together to create a vacuum that draws milk from the teats, alternating with air, which causes the milk to flow into a container in a rhythmic fashion (called *pulsation*), similar to a kid nursing.

Both goat and cow milking machines used to be not only expensive but loud and bulky. Newer ones such as the Simple PULSE (`https://simplepulse.com`) are lighter, quieter, and were designed with goats in mind. (The people who made it have a small goat farm in Oregon.)

Like any skill, machine-milking takes practice. Communicate with sellers of milk machines or try to find other goat owners who machine-milk so you can get an idea of what to expect from a milk machine and what using one entails.

Here are the steps for milking a goat with a milking machine:

1. **Get the goat onto the milk stand and secure her in the stanchion with some grain for her to eat.**

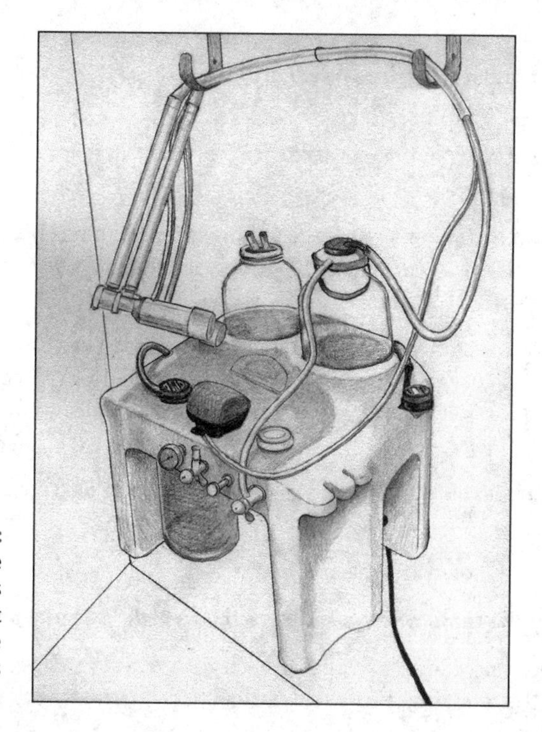

FIGURE 15-2: A Simple Pulse machine works well if you don't have a large number of goats to milk.

2. **Wash your hands.**

   You may need to rewash your hands if they get dirty or between goats.

3. **Brush off any debris on the goat and then wash and dry the teats and the area right in front of them.**

   Use a clean, soapy cloth or wipe, such as Milk Check Teat Wipes, for washing and a clean one for drying. Make sure the teats are completely dry.

4. **Hand-milk one or two squirts from each teat into a cup or towel.**

   This process, called *fore-stripping,* allows you to check for abnormalities and removes any milk close to the surface of the teat that is more likely to be contaminated with bacteria. If the milk is abnormal (with clumps, strings, pus, or blood), dispose of it after milking.

5. **Turn on your milking machine.**

   Turn the vacuum pump on and make sure that the pulsator is clicking and the vacuum pressure is set correctly before you attach the inflations to your goat. Also make sure that the vacuum pressure is set to the desired level.

6. **Attach the inflations to the goat.**

   Attach the inflations one at a time, making sure the teat is straight and milk flows smoothly. After both are attached, double-check the vacuum pressure reading and adjust as needed to the recommended pressure.

7. **Massage the udder gently, as needed.**

   You may need to massage to make sure the goat "lets down" her milk, or if the milk slows down but her udder doesn't seem empty. When it slows to a trickle, you're done.

8. **When you can no longer see a large volume of milk going through the tubing, turn off the vacuum pump or pressure shutoff switch and remove the inflations one at a time.**

   Make sure you keep an eye on the quantity. Over-milking can cause mastitis in machine-milked goats. The udder should shrink as you milk it, except during the first few days after kidding, when it may be swollen from the hormones released for kidding.

9. **Hand-milk the last bit of milk to prevent disease and decreased milk production.**

   Milking the last bit of milk is called *stripping.*

10. **Dip or spray the teats with a sanitizer such as Fight Bac teat spray.**

    If you use dip cups, use a clean one for each goat to avoid cross-contamination. If you spray the teats, make sure to thoroughly coat each teat end with spray.

11. **Return the goat to the herd.**

    Make sure you have some fresh hay or alfalfa and fresh water available for the goat right after milking. She will eat and drink instead of lying down and exposing an open teat orifice to bacteria. The orifice gradually seals, protecting the goat from mastitis.

12. **Clean the milking machine according to manufacturer's instructions.**

Handling Milk to Keep It Clean and Fresh

After you milk your goats, you need to take a few steps to ensure that your milk is clean, that bacteria growth is minimized, and that the milk stays fresh as long as possible. In this section, I tell you how to strain, cool, and store fresh goat milk.

Straining the milk

If you're milking by hand, you need to immediately strain the milk before you use it (unless you'll be putting it through a cream separator or using it right away to make cheese or another dairy product that requires heating). Milk obtained through a machine will go directly into a container. Straining the milk removes hair, hay pieces, or any other foreign object that may inadvertently get into the milk if you're hand-milking. Straining also gives you the opportunity to take a second look at the milk, in case you couldn't see an abnormality in the initial squirts. (See the "Getting a Grip on Hand-Milking" section, earlier in this chapter.)

To strain your milk you need a strainer with a milk filter, which you can find in feed stores or goat supply catalogs. If you don't have a lot of milk — say, a few gallons — a mini stainless steel strainer will do nicely. I buy the large milk filters and cut them to fit in the mini strainer rather than buying the hard-to-find and more expensive mini milk filters. The mini strainer sits on the top of a half-gallon canning jar for straining and immediate cooling.

Cooling the milk

To retain the milk's flavor and minimize the growth of bacteria, you need to cool milk quickly. If you have a timer (or a better memory than I do) you can put the milk in the freezer and then take it out before it's frozen. If not, then go to your favorite camping supply department and get some of those little plastic liquid-filled ice cubes that can be refrozen, put your jar of milk in a bucket or other container, surround it with ice cubes and put the whole works in the refrigerator until it is chilled. Then take out the ice cubes, refreeze, and use them after the next milking.

Another option for cooling the milk, which I've used for many years, is a mini fridge in the barn. As soon as I'm done milking, I put the jar(s) in the refrigerator and they cool down quickly. The next day, I take the jars from the prior day and move them to the refrigerator in the garage, where my buyers can take them and leave money.

Storing the milk

The easiest way to store milk is in the refrigerator at 40°F or lower. I use half-gallon jars with plastic lids. If promptly chilled and stored at the right temperature, goat milk will last five days or more before it begins to develop a "goaty" taste (you know what I'm talking about when you taste it!) or starts turning sour.

TIP

If your milk starts to go bad, feed it to chickens, pigs, or dogs — they love it.

To preserve your milk by freezing, you need to do so right after straining. Pour it into a quart-sized resealable plastic bag, leaving a little headspace. Mark the date on the outside with a marker and freeze at least two of them in a container, in an upright position. When they're frozen, you can put them in a gallon bag for freezer storage. It will last for a year, but it's best after two or three months. To thaw a bag, place it in a clean, dry bowl in the refrigerator. Don't thaw in hot water — doing so would cause it to separate.

Other ways to store milk require some chemistry — you can ferment it by making yogurt or kefir, or you can turn it into cheese. The appendix gives you some recipes, and you can find many more on the Internet or in one of the many books on cheesemaking.

Staying Legal while Selling Milk

In 1924, the United States Public Health Service (USPHS) developed a model law called the Pasteurized Milk Ordinance (PMO) to stop milk-borne illness. The intent of the law was to require that only Grade "A" pasteurized milk could be sold to consumers and businesses. So far, all but four states have passed the law or a modified version of it.

Although selling anything other than Grade "A" pasteurized milk is illegal in 46 states, farmers can still sell raw milk legally in 30 states. The states that have legalized the sale of milk that's not from a grade A dairy or pasteurized have done so by passing additional laws or administrative rules.

In other cases, small milk producers have gone around the law by selling shares in their animals, and some state regulators have chosen not to intervene. The share plan, which has been adopted by some people with cows, is like the *community supported agriculture* (CSA) model, in which people invest in (buy a share of) a farm and get vegetables and fruits throughout the season — only in this case they get milk.

Even if you can't sell milk for human consumption, you may be able to sell it to people for their pigs, dogs, or orphaned animals. In every state but Michigan, goat owners can legally sell raw milk for animal consumption.

Legally selling milk products such as yogurt or cheese without being licensed as a dairy is another story. Most states prohibit sales of these items by small, unlicensed farmers, although in some states the authorities' policy is not to actively seek out people who are selling these products outside of the law.

If you plan to sell your milk or milk products on a small scale, first find out what hoops you need to jump through. You can find individual state laws and regulations regarding the sale of milk at `www.realmilk.com/milk-laws-1.html`.

» Understanding the meat goat market

» Using other parts of the goat

» Finding out where to market your goats

Chapter 16

Goat Meat: From Breeding to Selling and Beyond

People around the world have used goat meat as a staple in their diet for centuries. According to the U.S. Department of Agriculture, goats produce very lean meat. It has 25 percent fewer calories than chicken and one-third as much fat as beef.

Meat from mature goats is called *chevon*, and meat from kids is sometimes referred to as *cabrito*. Whatever you call it, it's a tasty alternative to other meats, something that more people are learning every day, as goat meat gradually catches on in the United States.

TIP

If you want to make money raising goats, raising them for meat is an efficient way to do so. More than 60 percent of the goats raised in the United States are for meat. Demand keeps growing, but most goat meat is still imported to the United States from New Zealand; small-scale meat goat farmers can help meet the need.

In this chapter, I tell you about some of the important considerations for selling goats for meat.

Getting the Basics of Raising Goats for Meat

Any goat meat is good for eating. You may eat goat meat as a way to deal with excess wethers or get food from a predominantly milk or fiber herd, but if you're raising goats for income from meat production, you want goats that are solid, well-muscled, and bred for meat. The most common meat goats are

- Boers
- Kikos
- Kinders
- Myotonic, or fainting goats
- Spanish, or feral goats

Breeders are cross-breeding some of these goats to each other in hopes of producing a better meat goat. Two of these crosses include the *TexMaster* — a cross between the Myotonic and the Boer developed at Onion Creek Ranch in Tennessee and the *Moneymaker* — a cross of Saanen, Nubian, and Boer created on the Copeland family farms in California. (Chapter 3 runs through goat breeds.)

Because meat goats are still relatively new to the United States market, there's still a lot of room for cross-breeding in a quest to find the ideal. Before you decide on a breed, look around to see what's happening in your part of the country and think about whether you might want to work with more than one breed.

Cross-breeding standard dairy goats to produce meat goats

Most people don't milk their goats for years at a stretch. Instead, they breed them every year and give them a two-month break from milking. (The end of a pregnancy is hard enough on a doe without taxing her further by milking.) This schedule can easily triple a herd in one short year. Smart breeders keep only the best bucks for breeding. The market for pack wethers is limited, and standard-sized dairy goat wethers and does are less likely to be in demand as pets when people can get cute Nigerians or Pygmies. You have to sell a lot of kids if you raise dairy goats, and you may have a tough time finding buyers.

One solution to this problem is to cross-breed your standard dairy goats with a meat goat, such as a Boer. The offspring of Boer goats crossed with Nubian, Alpine, or another dairy breed grow faster than the dairy breeds and appeal to people who want to buy goats for meat. Nubian and Alpine crossbreeds are the most popular in the market. If you want a smaller meat goat, you can buy or make a Kinder goat by crossing a Pygmy and a Nubian to create a miniature dual meat and milk goat.

Using crossbreeds also has the advantage of being less expensive for another reason: Purebred meat goats can be expensive, in some cases requiring that they be shipped from a long distance. If you breed crosses instead of purebreds you just have to get one buck with good genetics and use him on any number of does.

Using your dairy goats for meat

An alternative to using purebred or crossbred meat goats is to use excess dairy goats for meat. Particularly if you have a large goat such as the Nubian, you can produce animals that provide meat for sale or for your own use.

Some people don't like taking their kids to the auction to sell them because they consider the treatment they get there inhumane. Instead they slaughter the goats themselves and put them in the freezer. Kids make very tender meat, while *cull does* (does that you sell because they don't fit with your breeding program) are less tender and can be used for ground goat or sausage.

Selling Your Products

Demand for goat meat in the United States is rising, but more than half the goat meat consumed here comes from New Zealand and Australia. Plenty of room for a local market! This section gives you the information you need to start filling that gap by selling your own goat meat.

Identifying potential buyers

Start your goat-meat enterprise by finding out who your buyers might be. Demand exists for meat goats, as well as for goat meat.

You can make more money selling meat goats with good genetics behind them than you can selling meat. Those likely to be interested in buying meat goats include the following:

- **People interested in showing goats:** Children in 4-H or Future Farmers of America can raise a meat goat as a project, show him at the county fair, and then sell him to the highest bidder. Both adults and children buy goats for shows sanctioned by meat goat registries, such as the American Boer Goat Association (ABGA) or the American Meat Goat Association (AMGA). In some cases, you can show market wethers that aren't even purebred.

- **Owners of acreage:** Goats are now recognized as superior to using herbicides to control noxious weeds. Because meat goats are big (in relation to other goats) they can more efficiently browse large areas, helping with weed control.
- **Goat packers:** Meat goat or meat-goat-cross wethers are becoming more popular for use as pack animals because they're large and strong.
- **Other breeders:** If you have some good purebreds or a new meat breed that you're developing, you can sell those goats to others who want to get into the business or improve their herd.

TIP

You get the best prices when you sell goats that weigh 40 to 60 pounds. The prices continue to go down as the goats become heavier, with those that weigh from 80 to 100 pounds bringing the lowest price. If you have to sell your heavier goats, do so when no lighter goats are available.

Selling goat meat rather than selling those goats *on the hoof* (alive) gives you the advantage of being able to ensure that they are treated kindly and slaughtered humanely. On the other hand, it also requires more work on your part — whether you are coordinating all the necessary steps or getting qualified as a slaughter facility. You also may have a harder time finding customers, but after you do, if you can supply them with the quantity and quality of meat they want, you're in business.

Try the following avenues for finding buyers for your goat meat:

- **Local restaurants:** Restaurants that serve people from an ethnic group in which goat meat is a staple or people who are interested in less common foods may want goat meat for their menus.
- **Prisons:** Prisons sometimes buy meat from cull goats at discounted prices, ensuring that you can sell meat even from your less desirable animals.
- **Farmers' markets:** Farmers' markets are springing up in many communities and draw a wealth of customers hungry for organic produce, raw milk and cheeses, and meat from animals that were fed and treated well.

  Farmers' markets can be a particularly good place to sell goat meat — if you live near people who traditionally eat goat meat and the local laws don't prohibit its sale. But people who have traveled and tried goat meat in another country or have heard about it and just want to try something new are also potential customers.

TIP

Be enthusiastic about your product. If you can, give out samples to show people how good goat meat really tastes. Provide recipes, information on nutritional value, and cooking tips.

The downside to farmers' markets is that they're time-consuming — you have to travel, set up and break down your vendor booth, and stay put to sell your product. (Unless, of course, you want to hire someone else to do the work.) On the other hand, they give you a chance to get away from the farm and interact with people.

- **People in ethnic communities:** People from Middle Eastern, Indian, Pakistani, Hispanic, and African origin regularly eat goat meat and are often interested in buying one or two goats at a time for a celebration. (Table 16-1 lists holidays when goat meat is commonly eaten.)

  A study in 2005 by the Northwest Cooperative Development Association in Washington showed that almost three quarters of people who eat goat meat do so for cultural, traditional, or religious reasons.

TABLE 16-1 Holidays Popular for Goat Meat

Holiday	Religious/ Ethnic Group	Date	Type of Goat Meat
Easter	Christian	Late March or early April	Milk-fed kids
Cinco de Mayo	Hispanic	May 5	Suckling kids or large weaned kids
Mexican Independence Day	Mexican	September 16	Weaned to yearlings
Navratri	Hindu	Late September or early October	Weaned bucks or wether kids
Ramadan	Muslim	April or May	Goats weighing 45 to 120 pounds, with an ideal weight of 60 pounds
Eid al-Fitr (the end of Ramadan)	Muslim	April or May	Goats weighing 60 pounds
Eid al-Adha (Festival of Sacrifice)	Muslim	The date varies by locality	Yearlings, not wethers
Various	Caribbean	Usually August	Young, smelly bucks

TIP

To maximize your income you need to time your goat sales to make the most from the market. You can get the highest prices around Thanksgiving and Easter and generally in the fall. Demand is highest at these times because of the celebrations when people are more likely to eat goat meat. Summer is not a good time to sell your goats.

Selling goats

If you are directly selling your goats to a buyer for meat, you have several options: You can sell the whole animal (privately or at an auction) and leave slaughtering up to the customer, or you can have it slaughtered by a mobile butcher or licensed slaughterhouse. Unless you are licensed as a slaughter plant, you can't do the slaughtering.

Selling a goat for a customer to take home to slaughter

The easiest transaction is selling a goat to a customer to take home to slaughter. When a customer comes to a farm to buy a goat, make sure you have separated those goats that are for sale from the ones that aren't, so that you don't have customers trying to talk you into selling a goat that you don't want to sell.

You need to agree on a price for the goat in advance, unless you want to bargain with a customer. If you price the goat by the pound, make sure you have a scale so that you can weigh the goat in the presence of the customer.

Then make the transaction, write a receipt for the purchase, load the goat in the customer's vehicle, and you're done. If you have concerns about the goat's welfare, talk to the customer about care long before you allow him to take the goat, and do not allow him to transport the goat in an inhumane manner — for example, in the trunk of a car.

Letting a customer slaughter on your farm

One way to make sure your goats are slaughtered humanely is to let customers butcher the goats they buy on your farm. If you decide to go with this option, remember that the law prohibits you from helping or participating. (See "Legal considerations" later in this chapter.) Before you allow customers to slaughter a goat on your property

- Consider the impact of having strangers come to your farm. They can bring diseases with them, particularly if they have other livestock, and they might run around into places you'd rather they didn't.
- Evaluate your legal liability. The section "Legal considerations" runs through important points you need to think about before you let anyone slaughter an animal on your property.
- Make sure that they are experienced and know what they're doing. Ask whether they have slaughtered a goat before and how they usually do it.
- Clarify that they must humanely slaughter the goat.

» If you haven't picked out a specific goat, separate the goats you're selling from the others to prevent last-minute bartering over a goat you don't intend to sell.

» If you have a livestock guardian dog or other farm dog, make sure that it's restrained or locked up.

Find out in advance who will be coming to slaughter the goat. A friend of mine had some buyers come out to her farm every year and slaughter a goat for a celebration. In previous years just men had come, but one year they brought children to watch, which made her uncomfortable.

TIP

If you allow others to slaughter your goats on your farm, provide them a place away from your other goats to avoid traumatizing the goats. Be clear about who will deal with cleanup and any waste products, such as internal organs, and make sure you know your state law regarding such disposal.

You may need to provide the following:

» A place to hang the carcass to drain the blood

» Sanitized table for cutting up the carcass

» Fire pit or blow torch for searing hair

» Very sharp knife

» Hose for rinsing blood or washing internal organs

» Shelter from inclement weather

Hiring a mobile butcher or going to a slaughterhouse

You can have a customer come to your farm to pick out a goat that you then have slaughtered. After the customer selects his goat and pays you for it, you call a mobile butcher or take the goat to an agreed-upon slaughter facility. This method ensures that the goat is slaughtered according to accepted methods and nothing goes wrong because professionals are doing it.

You can have the customer pay you for the goat and the cost of slaughter and the butcher for the cost of cutting and wrapping at the time he picks up the meat. (For more on mobile slaughter and taking a goat to a slaughterhouse, see below.)

Livestock auctions

Selling goats at a livestock auction is easy, but you won't get the best prices there, and the auction gets a fee from your earnings (usually a percentage of the sale).

Prices fluctuate, as well, so if you decide to take your goats, do so when the auction has a specific day for selling goats, or one or two weeks before a major celebration. (Table 16-1 shows you times of celebrations.)

A side benefit to auctions is that you can also meet people who are interested in buying directly from you in the future. Make sure you have business cards ready to hand out.

Advertising

Marketing your meat goats is like marketing anything else. You need to keep at it and you need to advertise in a variety of venues in hopes of finding the one that pays off.

You'll find lots of advertising opportunities for meat goats. Some of these include:

- **Classified ads:** Breed-specific journals and thrifties usually provide classifieds for a small fee.
- **Goat shows:** Exhibiting at shows enables you to market to other goat producers and interested parties watching the show. You can also find out about market opportunities by talking to others who are showing their goats.
- **Bulletin boards:** If you live near a city with a large ethnic community that traditionally eats goat meat, advertise on bulletin boards in stores or other businesses that serve that community.

  After you sell a goat or two to families in the community, you may find that customers come to you through word of mouth. I don't even raise meat goats but am asked by acquaintances from time to time about selling goats for celebrations.
- **Feature articles:** Consider writing an article about your farm or asking the local agriculture paper to feature your farm.
- **Television news:** Find a reason to have the local television station feature your goats and business on the news. Newscasters regularly do animal features because they know that viewers love them. When one of my goats had quintuplets I notified both print and television media, who were happy to come out.
- **The Internet:** Craigslist, Yahoo! discussion groups, and various agricultural sites provide a venue for advertising goats for sale.
- **Business cards:** Make sure you have plenty of business cards; always carry some with you and leave them in stores, auction bulletin boards, and other businesses where interested people can take one. You never know when you might be chatting with someone in a store who expresses interest in purchasing goats or goat meat.

Legal considerations

You need to be aware of certain legal issues relating to selling meat goats and allowing related activities on your farm:

- **Illegal drug residues:** You need to know the *meat withdrawal times* — the length of time during which a drug can be found in meat after you give it to a goat. Under Food and Drug Administration (FDA) policy, you're liable for any illegal drug residues found in your goat products. Ask your veterinarian about the meat withdrawal time for any drug he prescribes or that you give to your goat.
- **Slaughter regulations:** Under USDA requirements you can slaughter your goats and sell the meat only if you are licensed as a slaughter plant. If you aren't licensed, you can use meat that you slaughter only for home consumption or give it away.
- **Landowner liability:** If someone gets hurt while on your property with your permission and you are benefitting from them being there, you may be liable. Liability is a concern if you allow a visitor to slaughter a goat that she's buying from you and something goes wrong.

  WARNING

  Make sure you're covered. Talk to your insurance agent about liability insurance that would apply to an accident that came about from a goat sale and slaughter.
- **Rental or lease agreement:** If you rent or lease your farm property, read your lease carefully to make sure you aren't violating any provisions by having other individuals slaughter goats on your farm.

Determining what to charge

The market depends on whether you are selling goats on the hoof or goat meat, as well as a lot of factors outside your control, from the number of people in your area who are selling goats to the popularity of goat and goat meat in your community. To find out what people are getting for live goats:

- Ask other breeders what they charge.
- Go to a livestock auction where goats are being sold and see what kinds of prices they bring.
- Talk to your local slaughterhouse. Ask the local butcher whether they buy goats or can refer you to other meat goat sellers.

» Contact a local broker or cooperative. You can find a broker or cooperative that buys meat goats through your meat goat association.

» Price goat meat that is being sold in farmers' markets or specialty stores.

TIP

If you're selling meat to a customer, you need a licensed butcher. Normally customers agree to purchase a goat at a specified price per pound hanging weight. You need to make sure they understand that the hanging weight is more than the weight of actual meat they will receive. You also need to be clear on who will pay for what and make sure they talk to the butcher about what cuts of meat they want. The butcher can let them know their options.

Using marketing terminology properly

If you plan to sell your goat meat, make sure that you use the appropriate terminology and don't mislead customers with untrue claims. Here's a rundown of important terms:

» **Organically certified** refers to meat from a goat that was raised according to the National Organic Program standards and certified by an accredited state or private agency. Among those standards are dictates that

- You do not use any dewormers or antibiotics in the goat and in its dam during the last third of pregnancy or lactation.
- Feed and bedding must be from certified organic sources.
- You may vaccinate your goats.

» **Natural** means that the goat meat is minimally processed using no artificial ingredients, coloring agents, or chemicals. The raw product isn't fundamentally changed.

» **Humanely raised and handled** meat is a newer category relating to animal welfare. It means the goats are from a farm that is enrolled in a private certification program such as Humane Farm Animal Care (`www.certifiedhumane.com`). Currently, such certification programs have no standards for goats.

TECHNICAL STUFF

Grass fed does not presently have a regulated meaning for marketing, but that may change in time.

Slaughtering Goats

If you plan to eat your own goat meat or sell it to other people, you have to figure out how you want to get the goat slaughtered. You can butcher your own goats for your meat supply, but you can't legally slaughter goats for customers, unless your operation is licensed as a slaughter plant.

If you plan to use your meat only for home consumption or to give to friends, doing it yourself is an inexpensive option. (See the section "Legal considerations.")

In this section I talk about your options for how to get your goats slaughtered and give you the pros and cons of each of them.

Doing it yourself

If you're a do-it-yourself person or a hunter who's used to butchering animals, you probably want to slaughter your own goats for a family meat supply. In addition to being less expensive, doing it yourself is easier on the goats because they don't have to deal with strangers or be hauled off to an unknown place and handled by people they don't know.

If you want to slaughter your own goats, talk to other people who raise and slaughter their own meat goats or contact your local county extension office for information. You can also find instructions and even videos on the Internet. The upcoming section "Humanely slaughtering goats" gives you pointers on the best way to do your own slaughter.

Using a licensed slaughter plant

If you're too squeamish to slaughter your own goats or just don't have the time or experience, you can hire a professional to do it. In some areas you can find a licensed mobile slaughter company to come to your farm, but more commonly you take the goats to a licensed slaughter plant.

One disadvantage of using a slaughter plant is that you lose control of how the goat is slaughtered. If you hire a mobile butcher, you will at least be there to supervise. But if you take your animal to a slaughter plant or just take him to an auction, you will have no idea whether he is being humanely handled and slaughtered.

You can find a licensed slaughter plant by looking in the phone book or asking other farmers or the local feed store who is available and who does a good job. At certain times of the year, such as in the fall, you may have difficulty getting an appointment because they are busy. In addition, you may also have difficulty finding a plant that slaughters goats, depending on where you live.

Hiring a mobile butcher

The next best option to having your goats slaughtered on your farm is to hire a mobile butcher. Mobile butchers are licensed to come to your farm and slaughter your animals and then take them to a butcher shop for cutting and wrapping. They aren't available in every area.

I hired a mobile butcher for a pig I raised. It's better for your animals because they don't get stressed out by being loaded up and driven to a strange place. Instead, they can stay home, where they are comfortable and won't get frightened.

A mobile slaughterer has a specially designed truck where he hangs the goat carcass to remove the blood. He takes all of the internal organs and waste, so you have no clean-up afterward. If you have more than one goat that you want slaughtered, you save money by having all of them slaughtered at once.

After slaughter, the mobile slaughterer takes your goat carcasses to the shop to cut and wrap the meat. You just pick up your packages and pay the butcher.

Humanely slaughtering goats

Muslims and Jews have age-old traditions for slaughtering goats and other animals. Both religions require that the animals not be made to suffer. Currently accepted standards for humane slaughter emanate from these traditions as well as an ethic of kindness. Here are some tips for humane slaughter:

- If a goat is being slaughtered at your farm, make sure that he is away from the other animals, so that you minimize the trauma to the others.
- Keep the environment quiet and calm.
- If you have more than one goat to go to the butcher, send them together so they will have the comfort of their herd for the trip.
- Make the transport environment as comfortable as possible to avoid stressing them.

- Use a knife whose blade is twice as long as the width of the goat's neck to prevent pain during the cut.
- Don't unnecessarily bend your goat's neck when cutting the throat.
- If you are artificially restraining the goat, make one smooth cut within 30 seconds of restraint.

Using All of the Goat

If you are particularly practical or environmentally conscious, you probably want to use parts of the goat that are left after the rest of it is turned into meat. In this section, I talk about three possible uses.

Hide tanning

After you've skinned a goat, throwing that lovely hide into the compost or garbage seems like a shame. You can make a pillow, a rug, a wine bag, a drum, or even an article of clothing from a goat skin. I remember a man who proudly wore his goat skin vest to our annual goat conference.

TIP

Tanning a hide isn't very hard, but it takes some time. For best results use a fresh goat skin, rather than one that has been salted or stored in the freezer.

You need the following supplies to tan a goat hide with the hair attached:

- Water
- Laundry detergent
- Bleach
- Pickling solution
- Large plastic or glass container (a plastic garbage can will work)
- Nails and hammer
- Coarse sandpaper
- Large stick or baseball bat

Preparing a pickling solution

Before you get ready to tan the hide, make a pickling solution. You can use this four or five times as long as you have a place to store it.

1. **Boil five gallons of water and pour into a plastic (nonreactive) container.**
2. **Add 2½ pounds of alum (ammonium aluminum sulfate or potassium aluminum sulfate).**

   You can find alum at a local pharmacy, farm supply store, or through a taxidermy supply catalog.
3. **Add 15 pounds of rock salt, stirring until dissolved.**
4. **Cool the solution or store for later use.**

Tanning the hide

To tan a hide, follow these steps:

1. **Scrape all leftover fat or meat off the hide while it's fresh.**

   Place the hide on a smooth board with the fur side down, and use the back edge of a knife, held almost flat, to remove the scraps. (This is called *scudding.*)
2. **Hand wash the hide in warm water with laundry detergent and some bleach.**
3. **Rinse the hide with clear, cold water, making sure to get all the soap and oil off.**
4. **Put the hide in your pickling solution so it's completely immersed and doesn't have any air pockets.**
5. **Let it soak for two to three weeks, stirring the hide in the mixture once or twice per day.**

   Two weeks' soaking time is adequate for hides from young goats. Overtanning is unlikely with this method, so it's better to oversoak than undersoak.
6. **When it's done soaking, remove the hide from the pickling solution and repeat Steps 2 and 3 to remove all of the salt.**
7. **Carefully stretch the hide and attach to the wall (or a frame, if you have one) with small nails so that the hair side faces the wall.**

   Make sure that you hang it in an area away from sunlight. A garage or barn wall works well. Let the hide dry for four to five days.
8. **Soften the dry hide by rubbing the non-hair side with coarse sandpaper and then folding the hide, non-hair side out, and beating it with a bat or heavy stick.**

   Refold the hide periodically to ensure that you beat all of the hide to soften it.

Animal feed

Many veterinarians recommend the BARF (bones and raw food) diet for dogs. When you slaughter your own goats you can give your dogs the organ meat, bones, and other parts of the goat that you aren't going to use. And it's basically free.

Just separate out the parts that you don't plan to use, package them in plastic bags, or use them with other meat, vegetables, and supplements to make dog food, and freeze until you're ready to use. This lets you provide bones and meat for your dogs little by little.

Using the organs for herd health check

TECHNICAL STUFF

Sometimes goat herds develop nutritional deficiencies that can't be diagnosed on a live animal. (See Chapter 11 on nutritional deficiencies.) An example of this is copper deficiency. The only way you can tell if your goat herd has copper deficiency or another mineral deficiency is by analyzing the liver. And you can't have the liver analyzed while the animal is still alive.

When you butcher a goat, you have the perfect opportunity to send a liver to the lab for a trace nutrient mineral test. Whether you learn that the goat had a mineral deficiency or perfect lab values for minerals, you receive valuable information regarding your feeding program. (See sidebar for more on a trace nutrient mineral test.)

You can also check the goat's stomach or intestines for internal parasites, the liver for liver flukes, or other organs for abnormalities, depending on your level of expertise in evaluating these problems.

RUNNING A TRACE NUTRIENT MINERAL TEST

The trace nutrient mineral test is used to diagnose mineral nutritional status in a goat. Liver is the preferred specimen for this test. It should be kept frozen or refrigerated after you obtain the sample and during shipping.

This test determines whether the goat had a deficiency or an excess level of the minerals cobalt, copper, iron, manganese, molybdenum, selenium, and zinc.

» Getting ready to show

» Showing your goats in person or virtually

Chapter 17
Showing Your Goats

Showing your goats is a great way to see how they compare to other goats of the same breed, to learn about what makes a good goat for the type you're raising, to teach kids (human children, that is) how to prepare and handle goats, and to just have a good time and get together with other goat aficionados. Another benefit to showing goats is that it gives you a chance to find other goats to purchase and to market your own goats to other people.

You can have the greatest goat in the world, but if you aren't prepared and your goat jumps around the ring, tries to lie down when she's supposed to stand, and acts skittish and confused, the judge is less likely to see all of the goat's wonderful attributes. By the same token, if you don't know how to act, you may also lose points because you won't be presenting your animal to her best advantage.

If you decide to show your goats, you'll do better and feel better knowing what the judge expects from you. Children in 4-H can learn showmanship at their meetings. If you aren't in 4-H, go watch some goat shows to learn, or even better, watch 4-H-ers competing for showmanship. Check the schedule for your local county or state fair to find out when such a competition will occur.

In this chapter, I tell you how to get your goats ready for showing and show them to their best advantage.

Finding Shows

You can find goat shows for the breed you're raising by

- » Talking to other goat owners
- » Joining or contacting your local or regional goat club
- » Checking the websites of the registry you register your goats with
- » Contacting your local or state fair board to find out when the fair is scheduled and get on its mailing list
- » Asking 4-H leaders about shows
- » Putting on your own show

REMEMBER

Make sure to check well in advance for shows to avoid missing registration deadlines. Many clubs have annual shows, so if you miss it, you have to wait a year to plan for the next one.

Nowadays, some organizations are finding ways to hold goat shows without requiring you or your goat to be there physically. This is where the virtual show comes in. The Miniature Dairy Goat Association (MDGA) has been doing them for years.

TIP

If you can't find a show, consider talking to other goat owners about holding your own actual or virtual show and ask the association where your goat is registered about the requirements to do so. At a minimum, you'll need to hire a judge who is trained and certified by the registry for your type of goat.

Preparing to Show Your Goat

You can't just take a dirty, hairy, wild goat from the farm to a show, or even photograph one in a way that will properly demonstrate his assets. You need to do some basic training and pretty your goat up a bit first — washing, clipping, and trimming his hooves. If you can, attend some shows before you show your own goats or look at photos of those goats who placed high in a virtual show.

To show your goat to his best advantage, you need to have the goat groomed according to the accepted standard for his breed. You can get general information about breed standards from the registry for your breed and specifics from the show registration information. And, of course, you need to have a clean (usually chain-link) collar (and, for a Nigerian Dwarf or Pygmy, a lead).

REMEMBER

Whether you show a meat goat, a fiber goat, or a dairy goat, make sure the goat is

- Healthy, with no obvious transmissible diseases such as soremouth, pinkeye, or ringworm.
- Well nourished and not too skinny or fat.
- Registered with the appropriate registry. In some cases, such as 4-H shows, goats don't have to be registered.
- Washed and brushed to remove excess hair and dirt.
- *Fitted* (prepared for presentation) according to the breed club or goat registry's requirements for that type of goat and the show rules.

After you find a show and register, you're ready to prepare.

Getting your goat show-ready

Before you can show your goat, you need to get her ready to show.

Clipping your goat

You don't need to clip fiber goats for a show because they're judged in part on their fiber. You may need to clip meat and dairy goats before showing. Read the show rules before you clip your goat for a show. Some shows prohibit clipping and others require it.

TIP

If you do need to clip your goats, do it two weeks before a show. (See Chapter 9 for information on basic clipping.) This allows any mistakes to grow out and makes the goat look better with a slight growth of hair. When a goat show is held in the early spring in a part of the country where the weather is still cold (known as a *fuzzy goat show*), you only need to do minimal clipping.

WARNING

Let the clippers cool down when they get hot and apply clipper oil from time to time to keep them lubricated. These steps help prevent you from burning or cutting your goat.

To clip a goat for a show, you need

- Clippers
- Clipper oil
- Scissors

» Brush

» A #50 surgical blade

Follow these steps to clip a goat for a fuzzy goat show:

1. **Secure the goat on a stanchion by tying her with a short rope to a fence or by having another person hold the rope.**
2. **Brush and clean the goat.**

   Brush the goat thoroughly, removing any down. Then clean the goat's coat with a waterless shampoo or use a few drops of mild dishwashing soap mixed with warm water in a spray bottle. Mist the coat and dry with a towel. Do this several times over the whole body until you can't see any new dirt on the towel.
3. **Clip with the grain of the hair, starting around the neck and cutting it shorter in the front.**
4. **Trim the hair with the grain from the front of the throat down to the shoulder and around the neck to the back.**

   Put less pressure on the clippers to remove decreasing amounts of hair as you move to the back. Lightly clip to remove any long hair on the brisket.
5. **Remove excess hair, clipping with the grain, between the elbow and knee on each leg.**
6. **Still clipping with the grain of the hair, remove excess hair on the flank and belly.**
7. **Lightly clip along the grain to remove any long hair from the *escutcheon* (the area between the rear legs) and the back of the thigh.**
8. **Clipping against the grain, trim the sides of the tail and leave a 2-inch brush on top.**

   Use your scissors to even the end of the brush.
9. **Working against the grain, remove the hair at the top of the hooves and under the dew claw.**
10. **Clip inside the ears.**

    If your goat has ear tattoos for identification, this will help the judge see the tattoos. Use this opportunity to check for tattoos and compare to the tattoos on the registration certificate.
11. **Clip off a doe's beard, cutting against the grain.**

    Leave the beard on a buck; it makes him look more masculine.

12. **Working against the grain, clip as much hair off your goat's face and ears as you can get.**

    You don't have to trim the bony part of the face because the hair is thin here already. Trim up the neck and chin. Trim the cheeks on each side, keeping them symmetrical. Trim the *poll* (the top of the head where the horns were).

13. **Shave the udder of a senior dairy doe with a #50 (surgical) blade.**

TIP

    Clip the udder when it's full so you're less likely to nick or cut it. You'll also do a better job on a full udder.

    Lift one leg at a time to clip the sides and back of the udder. Hold each teat between the thumb and two fingers to avoid nicking as you trim around it.

    If the *medial ligament* (the attachment in the back that runs between the two halves of the udder) has a deep cleft, leave a little hair on the bottom of that line so it isn't accentuated. If the medial ligament isn't strong, leave a small amount of hair on the bottom of each half of the udder.

    If the *foreudder attachment* (the attachment of the front of the udder by the belly) isn't smooth, leave about 2 inches of slightly longer hair down the middle of the belly to give it a smoother appearance.

14. **Brush the hair off your goat and evaluate the result of your clipping.**

    If necessary, neaten up by removing excess hair that you've missed. You can use scissors or press lightly with clippers, going with the grain of hair.

15. **Give your goat a final brushing to remove cut hair and then put her away.**

Trimming hooves

You need to trim your goat's hooves a day or two prior to a goat show. (Chapter 9 explains how to trim hooves.) Untrimmed hooves affect the way a goat walks and holds himself, which contribute to how a judge scores the goat. Some judges even lift up the goat's feet and look at the hooves.

WARNING

Take extra care not to cut too deep. You don't want your goat limping around the ring.

Practicing showing techniques

If a goat show is the first time your goat becomes acquainted with a collar and a lead, don't expect your goat to do very well. Even a well-trained goat may be a little nervous during the show because of unfamiliarity with the show ring, but if you lead-train first, at least the goat has an idea of what you expect. (To find out about lead training, see Chapter 8.) A goat with outstanding conformation who's leaping around the ring rather than walking slowly and calmly will most likely not place as high as she should.

The first time I took my goats to a show, I had no idea what to expect. My goats were totally out of control and placed fifth and seventh, to the chagrin of the person who sold them to me. By the next show, we had more experience; they knew how to walk and they placed first and second. This gives you an idea of how much of a difference experience walking with a goat can make.

TIP

While you're working with your goat on walking, take the time to practice setting her up the way that a judge would like to see her so she also has experience with that.

To set up your goat, stop walking, and then go to one side. Make sure the front legs are placed squarely under the goat's shoulders (not splayed out) and her back legs are evenly aligned with her hips. Stand up and make sure the goat's back is straight. Hold up the lead or collar so your goat holds her head high.

Even when you aren't being judged on your showmanship abilities, knowing how to set up your goat properly shows your goat to its best advantage and can contribute to a higher placing in the ring. Setting up is critical for a virtual show because the judge will be looking at that alone, without the advantage of seeing how the goat walks.

Assembling your supplies

If you're going to a live goat show, you need some supplies. The supplies you need to bring depend on whether the show is part of a fair that requires animals to remain for a number of days for exhibit or is just a local one-day goat show. At some shows, animals simply stay in their transport vehicles instead of moving into temporary pens.

Some basic supplies you need to bring to a show include

- **Feeding supplies:** Water buckets, feed bowls, hay racks or bags, hay, grain
- **Cleanup supplies:** Bleach water or another sanitizer, spray bottle, broom, pitchfork
- **Grooming supplies:** Clippers, hoof trimmers, stanchion, soap and towels, coveralls, brush (for last-minute preparation)
- **Showing supplies:** Show clothes, show collars, leads, registration papers, show program
- **Marketing materials:** Farm sign, sales list, educational brochures

- **Miscellaneous:** Bedding (if not supplied by the organization holding the show), milking supplies, short hose (to attach to a water supply), health certificates (if required), camera (to photograph winners), chair, goat coats (if overnight or cold weather), first-aid kit

Earning more than just ribbons

Most goat shows don't make money for goat owners; they cost money to enter. The exception is state or county fairs, which cost to enter but also pay premiums when your goats win. The amounts for premiums vary from fair to fair, but if your goats place high, you can make some money to help with your feed bill.

Fairs also help you make money on your goats indirectly by giving you a venue to display them in. You get a chance to promote your goats and educate the many people who come by. If you have ribbons to display, visitors may be even more interested in buying some. In some cases, people I talked to at a fair bought goats from me several years later. (See the next section for more on how you can attract potential buyers and sell your goats.)

If your goats place well in the competition, other breeders who are showing their goats or just watching the show may be interested in buying one of your herd. On the other hand, if you're like most of us goat lovers, you may come home from the fair with one more — instead of one fewer — goat.

Marketing at shows

TIP

Showing your goats is marketing. You're getting them out in front of potential buyers, even if you aren't aware of it. That's just one more reason to exhibit good showmanship and show off your animals to their advantage.

Your goat pens or exhibit area also can help your marketing efforts. You can get a relatively inexpensive sign or banner to attach to pens so people know who your goats belong to. Make sure to keep your area neat, and include marketing materials there and at any information tables provided. Ideas for marketing materials include

- Business cards
- Sales lists of goats you currently have on the market
- Educational materials, such as brochures describing the breed of goat you raise or information on basic goat care
- Photos of your goats and information on awards they've won

REMEMBER

Talk to other exhibitors and any visitors when they're looking at your goats. Politely answer questions and don't be shy about letting them know about goats you're selling and why they're a good buy.

Showing Your Goat in Person

As the show time approaches, you'll have practiced walking with your clean, well-groomed goat and setting him up. There are some specific requirements that judges are looking for in addition to the goat himself. From the time you enter the show ring until the time you leave, keep top of mind that you're showing your goat — first impressions count.

Dressing appropriately

Most goat shows require you to wear certain clothes when showing your goats. The standard for dairy goats is a white shirt and either a white skirt/pants or black pants. Many people wear a white shirt with their herd name and logo on it. (You can have these made by a company that does embroidery or screen printing.)

If you're showing meat goats, you may not be required to wear any specific colors — you may even be allowed to wear jeans — but it is expected that the person showing the goat be neat and clean. North Carolina Extension recommends a button-down or polo shirt and a belt.

You generally won't be barred or removed from a show for wearing other clothes than those recommended, but if you're being judged on showmanship, you'll lose points.

WARNING

Don't wear flip-flops or soft shoes in the ring in case a goat steps on your foot. Try to keep your show clothes clean before you enter the ring.

Keeping your goat between yourself and the judge

REMEMBER

When you show your goat, you want the judge to be able to see the goat at all times. Never get between your goat and the judge.

To change the side of the goat you're walking next to, do a roll turn. If you're walking clockwise with the lead in your right hand, just before you get to the judge, turn around the front of your goat and change to your left hand. You can reach over your goat to set the legs that are on the far side from you. Always set the legs closest to the judge first; if the judge is in front of you, for example, set the front legs first.

Move to the front of your goat when the judge views the goat from the rear. As the judge moves to the right side and around to the front of the goat, move to the goat's left side, keeping him between you and the judge. When the judge is in front of the goat, remain on the goat's left side, so the judge can see the front view. Hold the goat's head up with the collar, your left hand under the goat's jaw. As the judge moves to the left of the goat, move back to the front of the goat so the judge can see the entire animal.

Focusing on the judge at all times

The judge will tell you what you need to do and may ask questions about your goat. If you're distracted, you may lose track of what's going on. Keep your eyes on the judge so that you always know what's happening.

Even though the judges are trained on the same standards for evaluating the goats, you'll find some variations. Keeping your eyes on the judge lets you know what to do next and may even give you clues about what parts of the goat the judge is most interested in.

Making sure not to talk with your neighbor

Talking with the person next to you while the judge is thinking about placements, comparing goats, or examining each goat is considered rude, distracting, and unshowmanlike.

REMEMBER

You're in the ring to show your goat to the best of her advantage and focus on the task at hand. The judge isn't supposed to judge the goat based on the handler's behavior, but she has the power to place your goat anywhere.

Remaining calm even if your goat is misbehaving

Despite being trained and walking on a lead perfectly at home, a goat may get bored, scared, or just ornery and misbehave in the show ring. Don't ever hit the goat or treat him roughly. Remain as calm as possible and do what the judge asks to the best of your ability. Some goats act up from time to time, so you just have to maintain a sense of humor and a good attitude.

Doing what the judge asks

You'll start by walking the goat into the ring in a clockwise direction. Keep your goat under control with her head up, and keep her between you and the judge. If your goat doesn't want to move, gently lift her tail. When the judge asks you to stop, set your goat up in a straight line, facing the same direction as the other goats. Stand or squat on the goat's other side or near her head (see Figure 17-1). Stay alert for another request and keep the goat set up.

FIGURE 17-1: Setting up a goat.

If you're asked to change your goat's place with another goat, turn the goat around and move behind the line of other goats to that position. If you're asked to walk your goat to compare with another goat, walk your goat slowly, at the same pace and as near to the other goat as possible for easy comparison.

Keeping your goat properly set up

When you're showing a goat, the judge will ask you to stop and line up so that all the goats face the same direction. Then she'll evaluate each goat based on the criteria for the breed she's judging. You need to make sure you're showing your goat to his best advantage.

To set up your goat make sure that his weight is evenly distributed. Don't splay the legs. Set up a dairy doe so that you can see a third of the udder in front and a third behind the back leg. Set up the rear legs first and then the front legs. Hold the goat's head up. Make sure her front legs are straight down from her shoulders. (Figure 17-1 shows a girl setting up a goat.)

During a show, continue to keep your goat set up and show her to her best advantage even when the judge is focusing on another goat and even if you think the judge has determined the goats' placements.

If the judge asks you to put your goat in a line with other goats, keep your goat's shoulder lined up with the first goat in the line and parallel with the other goats. If you're lined up head to tail, try to keep a straight line with the goat ahead of you.

A judge can change her mind at any time during the show. You can relax a little, especially if your goat is getting upset about standing so long, but continue to keep your animal in place and looking as good as possible until the judge dismisses the class.

Being a good loser (or winner)

After judges determine rankings, they explain their reasons for awarding placements as they did. Listen closely to those reasons. Regardless of your placing, congratulate the class winners and other goat owners who placed ahead of you. Make an effort to learn from the experience. If you lose, don't complain. If you win, be gracious.

You may ask the judge about placements after the show is over if you truly need help understanding the judge's reasons. Don't question the judge in the ring or delay the next group of goats to be shown by asking questions.

Showing Your Goat Virtually

Virtual shows (also known as *V shows*), though not as fun as live shows because you can't get together with other goat fanatics, do provide a way to show goats while staying away from crowds. Unfortunately, they aren't currently offered by most registries.

Where they are available, they give you an opportunity to show off your goats and add credibility regarding their superiority when you want to sell them, if they place well. They also get your goats used to being set up and help prepare you for when you can attend a live show. In addition, they save you and your goats the hassle of getting to a show and being exposed to outside goats.

Setting the scene

Because the focus is on the goat being photographed, you need to find an uncluttered area on which to work. The ground should be flat, without hay or grass covering any of the goat's feet. A surface like wood or concrete is best. This ensures that the judge will be able to see the foot and leg areas, which are part of the scorecard.

As in a live goat show, make sure the person showing the goat is neat and clean, without distracting clothing or a hat.

The background should also be simple. If you don't have a plain wall that will work, consider hanging a sheet so that the lines of the goat's body stand out clearly.

Preparing your goat

Clip your goat so that the photos can reflect the conformation and so that the udder and legs show up well. If you're showing a doe in milk, make sure not to milk her until after the photo is taken. Also make sure that the udder hair is completely clipped so that all aspects of it can be clearly seen.

You need a handler who knows how to properly set the animal up while photos are taken. That person shouldn't be the focus of the shots, but he should blend into the background as much as possible. He needs to be able to set up the goat (see "Showing Your Goat in Person," earlier in this chapter).

Photographing properly

In general, you'll need photos of the goat from the front, the side, the rear, above, and behind. For does, because the udder is part of the scoring, you'll also need a photo of the fore udder.

Photos should be high resolution so they're as clear as possible. One nice thing about digital cameras is that you can take as many pictures as needed until you get the perfect one.

TIP

According to the MDGA, there are a number of things to remember when photographing a goat. The MDGA recommends that you not take pictures on a bright, sunny day, in order to avoid shadows. It also suggests that pictures be taken from the goat's level and "square on" and not too far away. For specific photos, the MDGA suggests the following:

- **Front:** The front legs need to be straight under the goat. The focus is the width of the chest.
- **Side:** Remember to keep the head up, the front legs straight under the goat, and the rear hocks under the pin bones.
- **Rear:** The focus for the rear photo is the escutcheon. The ideal escutcheon is high and wide. The feet and legs should also show well.
- **Above:** Judges are looking for a straight line from head to tail, so make sure the body is straight and head up.

IN THIS CHAPTER

» Creating miniature breeds

» Backpacking with goats

» Putting your bucks to work

Chapter 18

More Benefits of Goats: Fiber, Breeding, Weed Control, and More

Goats offer so many opportunities to enrich your life and produce additional income, products, and satisfaction. Besides using them for milk and meat, or for pets, you can raise them to produce fiber for sale or for your own use. And regardless of what kind of goats you choose, you can involve them in different hobbies and recreational activities — such as showing, creating a new breed, or just hanging out to relieve stress.

In this chapter I tell you about raising goats for fiber production, what fibers different goats produce, and how to make goat-hair fiber into yarn that you can knit with or take to market. I also tell you about caring for the up-and-coming mini dairy goats and about some of the other ways that your goat herd can pay its way or help you get more joy out of life.

Harvesting and Selling Fiber

Humans have kept goats and used the fiber they produce for millennia. You find goat fiber in clothing and in fabrics for furniture upholstery, drapes, rugs, and even wigs. Cashmere sweaters, made from the undercoat of certain kinds of goats, are highly prized. And they should be: Just one cashmere sweater requires a year's worth of growth from three to four goats.

Depending on the goats you choose and the type of fiber they produce, you can harvest fiber by shearing or by hand-pulling. Then you have to clean it, separate it, and go through a few other steps to get in shape for spinning.

Many people who raise small herds of goats for their fiber handle all these steps themselves, and they form communities so that they can get together or share their knowledge with each other.

In this section I talk about the different kinds of fiber goats produce and how to harvest it, process it, use it for your own projects, or sell it.

Reviewing fiber types

Different goats produce different types of fiber, some more valuable than others. The three most common types of goat fiber are the following:

- **Mohair:** Angora goats produce brown, black, red, and white fiber called *mohair.* (Angora fiber is produced by rabbits.) Their fleece grows about an inch each month, and you need to shear them twice a year, usually in the spring and fall. You can expect to get an average of 5 to 10 pounds of mohair from each shearing.
- **Cashmere:** The fiber that comes from any type of goat other than an Angora is called *cashmere.* It's the undercoat that you see on goats in varying degrees at the end of winter. Certain goats are bred for cashmere because they can produce up to 6 ounces of cashmere each year. The cashmere from solid-colored goats is more valuable than that from mixed-colored goats. In order to be considered cashmere, the undercoat has to be 19.5 microns in diameter or less.

  Cashmere-producing goats have both *guard hair* (long hairs forming the outer coat) and *downy fiber* (the softer undercoat), which need to be separated before spinning. You can do this by hand, but it takes a lot of time.
- **Cashgora:** Cashmere goat breeders use the term *cashgora* to refer to fiber from cashmere goats that doesn't meet the criteria for cashmere. Some goats can have fiber that is deemed cashgora at one shearing and cashmere at another.

Fiber also commonly comes from two miniature fiber goats. *Pygora* goats are a cross between an Angora and a Pygmy, and *Nigoras* are a cross between an Angora and a Nigerian Dwarf. You might think of a Pygora or a Nigora as a sort of mini Angora. They can produce three different types of fiber, or different variations and combinations among these types:

- **Type A** is similar to mohair.
- **Type B** is like a blend of mohair and pygmy goat fiber.
- **Type C** is similar to cashmere.

You can expect a Pygora or a Nigora to produce from 6 ounces to 2 pounds of fiber at each shearing. You can shear goats that produce type A or A/Bs twice a year. You don't even need to shear goats with type B and C fiber because they shed their downy undercoats in the spring, which means that you can comb or pluck the fiber instead.

WARNING

All types of fiber mat if they are left on the goat too long. The upcoming section "Shearing: How and when" lets you know when the time is right for removing the fiber.

Like cashmere-producing goats, most Pygoras and Nigoras produce guard hair and cashmere fiber, which need to be separated before spinning in order to create a better yarn.

Shearing: How and when

You need to shear Angora goats twice a year to get the most fiber. You don't shear cashmere- or cashgora-producing goats; instead you comb or pluck the fiber once a year.

WARNING

You need to shear your mohair-producing goats in the early spring and early fall. Make sure not to shear cashmere-producing goats, because you will lower the value and quality of the fiber by mixing coarser guard hair with the fine, valuable cashmere.

Shearing isn't hard to do, but it's hard on your back because you have to bend over (see Figure 18-1). If you have only a few goats you can use scissors or hand shears. If you can afford it and want to do all the shearing yourself, you can also invest in electric sheep shears, which range from $300 and up. Otherwise, you're better off hiring a professional shearer to come to your farm. Professional shearers charge a per-goat fee.

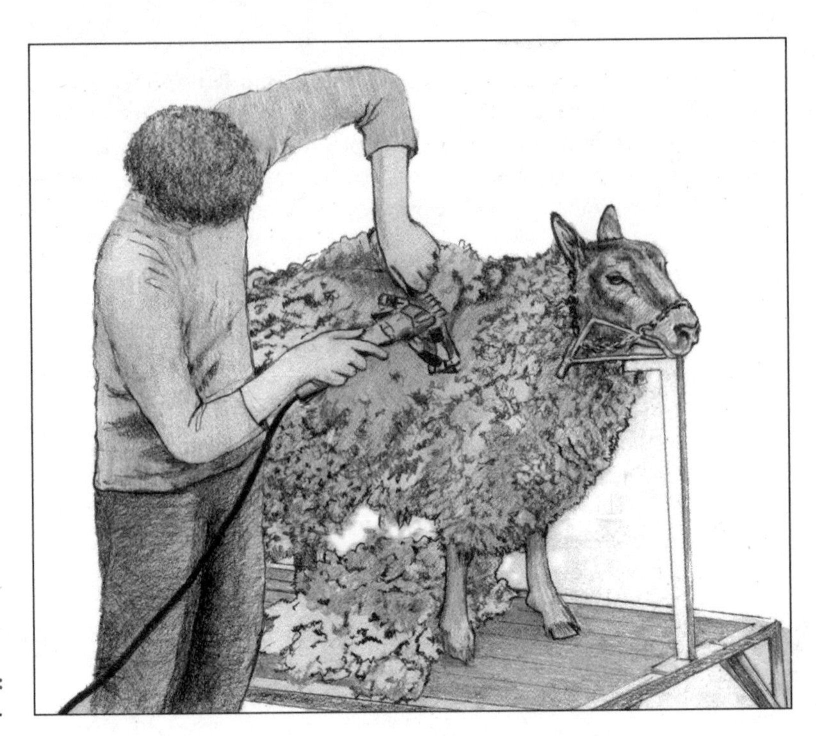

FIGURE 18-1: Shearing a goat.

You can work in tandem with the shearer — assisting, sorting, and bagging the fleeces while she shears. You can also give your goats needed vaccinations (see Chapter 10), deworm them, and trim their hooves at the same time you're having them shorn, saving you effort and preventing them the stress of getting out and being restrained twice.

To prepare your goats for shearing:

- A few weeks prior to shearing, use a pour-on insecticide containing permethrin or pyrethrin to kill lice and ticks.
- If the weather is rainy or snowy, keep your goats confined for 24 hours before shearing so that they stay dry.
- Clean and add new bedding to a dry shelter to keep the goats out of inclement weather for a month or so after they have been shorn. They're more prone to health problems such as pneumonia without their protective coats.

Start your shearing with the youngest goats and work in order of age because the youngest usually have the best fiber.

TIP

When you're shearing a goat, always use long, smooth strokes. Doing so keeps the fleece in longer pieces, which makes it easier to work with and increases its value. Be careful not to cut the skin, taking special care on the belly, the area where the legs and body meet, the scrotum, and the teats. If you do accidentally cut a goat while shearing, treat it with an antibiotic spray such as Blu-Kote.

Before you shear, get the following supplies together:

- Blow dryer
- Sheep shears
- Grooming stand or stanchion
- Scissors
- Paper bags, pillowcases, or baskets to hold the fiber
- Postal or hanging scale for weighing fiber

Follow these steps to shear a goat:

1. **Secure your goat on the stanchion or grooming stand.**
2. **Blow any hay or other debris out of the goat's coat.**

   Use your blow dryer on high speed.
3. **Shear the goat's belly.**

   Start at the bottom of its chest and move to its udder or scrotal area.
4. **Shear each side.**

   Work from the belly up to the spine, back leg to front leg.
5. **Shear each back leg.**

   Work from the beginning of the coat upward to the spine.
6. **Shear the neck.**

   Start at the bottom of the throat and work to the top of the chest on the bottom and from the chest to the ears on the top and sides.
7. **Shear the top of the back.**

   Work from the crown of the head to the tail.
8. **Remove any excess hair that you missed with your scissors.**

   One area often missed is in the area of the udder or testicles.
9. **Release your goat.**

10. **Check the fiber.**

    Separate any stained or soiled fleece or other contaminants. Weigh the unsoiled fleece, roll it up, put it in a paper bag, and mark the bag with the weight of the fleece, the goat's name and age, and the date sheared. Store fleece in a dry area.

11. **Sweep the area.**

    Make sure the next goat to be sheared starts with a clean area.

REMEMBER

Mohair and cashmere fibers are harvested differently. You comb or pull cashmere from the goat while you shear mohair. If you shear the cashmere, you'll wind up with a lower quality product because it still contains the coarse guard hairs. Because of these longer, coarser hairs, you need to *dehair* (remove the hairs by machine or by hand) cashmere before processing.

Processing the fiber

Before you turn your goat fiber into yarn, you need to put it through several processes. To process your fiber, you need to go through the steps in the upcoming sections.

Washing

Unless you're selling your raw fleece to a commercial operation that purchases fiber from many buyers, you need to wash your sheared or plucked goat fleece to remove grease, dirt, and other impurities. To wash the fleece, you need:

- Mesh bags
- pH test strips
- Baking soda
- Hand-washing detergent, such as Calgon
- Strong rubber gloves

Follow these steps to wash your goat fleece:

1. **Separate the fleece into smaller locks (bundles) and put them into mesh bags.**

    Do not pack the fleece tightly into the bags.

2. **Prepare your washing solution.**

   Put 145°F water in your sink or another container. Wearing rubber gloves, check the water's pH. If it is below 8, thoroughly mix in small amounts of baking soda until the pH is 8 or 9. (Neutral pH is 7.) Then mix in a small amount of detergent. This combination helps remove any oils or dirt.

3. **Wash your fleece.**

   Place your bags of fleece in the water and soak for 15 minutes, checking the water to make sure it remains higher than 125° Fahrenheit.

   Pull out a bag and check a lock of fleece. If it is gummy, all of the grease is not out and you need to keep soaking it for up to another 45 minutes.

4. **Wash the fleece a second time.**

   Refill the sink with 145°F water, using only half the detergent and no baking soda. Soak for 15 more minutes.

5. **Rinse the fleece.**

   After the grease is out of the fleece, remove the fleece from the water and let out the water. Fill the sink with more hot water and soak the fleece for 15 to 30 minutes, agitating it with your hand from time to time.

6. **Rinse again.**

   Remove the bags, refill the sink, and rinse for 15 to 30 minutes, using cooler water each time.

7. **Rinse a final time.**

   Check the pH of the final rinse; it should be 6. If it is higher, add a small amount of vinegar to lower the pH. A high pH can make the fiber less downy. Soak and rinse for 15 minutes.

8. **Dry the clean fleece.**

   Remove the bags, pressing gently, and then remove the mohair from the bags. Spread out the fleece to dry on a towel as you would a wool sweater. You can use a fan to accelerate drying.

Dyeing

You can use or sell your fiber in its natural color, or you can dye it. If you want to dye your fiber, do so after you have washed and dried it. You can use a commercial dye, a natural dye, or even Kool-Aid. Goat fibers take dye well but they have to be clean or the color will turn out uneven. Dyeing can coarsen cashmere somewhat.

TIP

You can get more information and tips on dyeing, using natural dyes, and books on dyeing at www.allfiberarts.com/cs/dyes.htm.

Carding or combing

Carding and *combing* are ways to separate the strands of washed fiber to prepare it for spinning into yarn. Both methods blend the fibers and remove hay and other contaminants that may still be left in the fiber. They also straighten the fibers to make them lie in the same direction. Carding produces a fluffier end product than combing because combing better aligns the fibers and makes the fleece more compact.

TECHNICAL STUFF

The end product of carding or combing is a clump of fiber with individual fibers aligned called a *rolag, roving,* or *batt.* You create a rolag when you roll the finished fiber off the carder. When you form it into a long narrow bundle it is called a roving. A batt is simply a clump of carded or combed fiber. You use a rolag or a roving to spin. You use a batt (think of cotton batting) to stuff pillows and other items.

You can card your fiber by hand or with a carding machine. You machine-card with a *drum carder* or hand card with two *hand carders.* A drum carder is usually a hand-cranked machine with two toothed rollers that card the fiber. Hand carders are square or rectangular paddles that look a bit like animal grooming brushes with small wire bristles or pins.

To comb your fiber you can use paddle combs. These are similar to hand carders, but they are simpler and have only one or two rows of teeth. They are good for working medium to long fibers or for working with cashmere, where you have to remove the guard hairs as you go along. Combing separates the long and short fibers and prepares the fiber for spinning.

WARNING

Combs are sharp, so be careful when handling them and don't leave them where children can get them.

You can find combs or carders at spinning supply stores or at an online specialty shop such as Pacific Wool and Fiber (www.pacificwoolandfiber.com) or woolcombs.com (www.woolcombs.com). You can find a number of videos that demonstrate carding and combing on YouTube.

Spinning

Spinning is simply using a tool to twist fiber to make yarn. Hand-spinning is the traditional method and requires a *drop spindle* (a weighted tool that you spin by hand to create yarn) or a *spinning wheel* (a machine with a wheel that you spin by using your feet). Commercial yarn producers use larger machines to produce quantities of yarn.

TIP

If you are raising fiber goats and want to get into spinning, I recommend starting with a drop spindle before purchasing a spinning wheel. You can make your own or buy one for a reasonable price. You can also learn more about using a drop spindle or a spinning wheel from a spinning book at `www.joyofhandspinning.com` or `www.handspinning.com`.

You need to practice spinning the drop spindle without yarn to get used to working with it. To spin with a drop spindle you attach a piece of yarn (called a *leader cord*) to your spindle and then attach the end of your fiber to that. You have to spin in the same direction (normally clockwise) to hold your fiber together. As you spin the wheel, you gradually add fiber to the end of the rolag or roving you are working with.

Spinning on a wheel requires you to use your hands and feet. You use your feet by pushing on a treadle that keeps the wheel moving as you gradually add fiber to be spun into yarn. You need to practice to learn to move your hands and feet together and get a consistent product. After you get it, the process seems like second nature and spinners even say it's incredibly relaxing.

Spinning wheels are expensive, and a variety of types are available. Before you run out and buy one, you need to answer several questions.

- Do you want to spin thin or thick yarn?
- Do you want a single or double treadle machine?
- How much money can you afford?
- How much space do you have to work in?
- Is the machine upgradeable so you can modify it to produce a different type of yarn?

Talk to other spinners about what type and brand of wheel they use and what they recommend. Try out several spinning wheels before buying one if you can. If you plan to buy a used one, make sure it is in good working order before buying it.

Selling your fiber

If you have more fiber than you can use, or if you don't want to use the fiber yourself, you can sell it to other textile producers or spinners after you have washed it.

The markets for mohair, cashmere, and cashgora vary from year to year. Prices can be as low as $3 a pound and as high as $12 a pound for white mohair. You can't usually sell colored mohair commercially but you can get more than $25 a pound by selling to hand-spinners. The fiber from younger goats is more valuable because it's finer.

BLACK SHEEP GATHERING

Every year in Lane County, Oregon, an event called the Black Sheep Gathering (www.blacksheepgathering.org) brings together people who raise colored fiber sheep, Angora goats, and Angora rabbits. The Gathering holds a large trade show with fiber supplies; provides educational information and demonstrations; sells hand-spinning fleeces; holds a spinners' circle; and holds goat and sheep shows. It's a great place to network with other fiber goat breeders and learn the ins and outs of producing yarn from your goats' fleece.

Markets for fiber abound. Some ideas for selling your fiber include

- Farmers' markets
- Online sites such as eBay, Craigslist, or Etsy
- A *cooperative* (an organization where fiber goat raisers join together to sell their fiber in bulk at a fair market price) such as Cashmere America Cooperative
- Conventions such as the Black Sheep Gathering (see sidebar)
- Advertising on a website or in local "thrifty" newspapers

Creating Mini Breeds

For some reason, people love miniature animals. My first goats were Nigerian Dwarves because I loved their small size, their markings, and the fact that they produce such good milk, which has high butterfat and protein and makes a proportionately large amount of cheese compared to some other dairy goats' milk.

While I was proud of my Nigerian Dwarves, I soon became enamored with the Oberhasli because of its beautiful red and black markings. In addition, although I was breeding my goats for milkable udders and had started with a champion Nigerian, I liked the idea of creating a new breed and increasing my goats' milk yield, as well as making some money. Urban farmers and people with small acreages are more likely to buy miniature, yet functional, goats than the larger breeds because they are easier to handle, take up less space, and may even be the only kind of goat legally allowed. (For more on choosing the right goats for your needs, see Chapter 7.)

MY MINI-TALE OF MINI-BREEDING

I started my mini Oberhaslis (I call them Oberians) by breeding two Oberhaslis to two bucks — one black and one buckskin (tan with a black cape and facial stripes). I got lucky and got correctly marked first generation (F1) kids, but found that in future generations that wasn't always the case; some had white spots that aren't correct for the breed standard. The minis are about the size of a large dog. Minimum height for does is 21 inches and for bucks it is 23 inches.

They can produce more than 8 pounds (1 gallon) a day, but I've found that my average Oberian produces from 1 to 2 quart total, when milked twice a day. They also eat only two-thirds as much as a standard breed.

When you breed mini-goats, you create a new experimental breed. The Pygora is a miniature fiber-producing goat from a cross between the Angora and the Pygmy, and the Kinder is a miniature milk and meat goat from a cross between a Nubian and a Pygmy. When these breeds were introduced, the Nigerian Dwarf was still considered an exotic breed and was more expensive and harder to get.

Now Nigerian Dwarves are quite popular and easy to find. As a result, hybrids using the Nigerian Dwarf are appearing.

- Mini dairy goats are created by crossing a Nigerian Dwarf with any of the standard breeds of dairy goat (see Chapter 3). Their breed standards mirror those of the full-size goats, except that they are expected to be smaller.
- Mini silkies are another type of goat that is growing in popularity. These beautiful little goats come from a cross between a Nigerian Dwarf and a Fainting goat. They don't have to faint because they are instead valued for their long, silky coats.
- Nigoras are miniature Angoras that are bred using the Nigerian Dwarf rather than the Pygmy. They produce colorful fiber similar in quality to that of the Pygora.

You can breed minis without registering them, and many people who want miniature goats don't care about having registered goats or are even opposed to registering their goats. But consider registering or buying registered minis, because you can't usually go back and register unregistered goats and their offspring.

TIP

If you're interested in raising mini goats, contact one of the mini goat associations for the breed you're interested in to learn more about them:

- **American Nigora Goat Breeders Association:** `http://nigoragoats.homestead.com`
- **Kinder Goat Breeders Association:** `https://kindergoatbreeders.com`
- **Miniature Dairy Goat Association:** `https://miniaturedairygoats.net`
- **The Miniature Goat Registry:** `www.tmgronline.com`
- **Miniature Silky Fainting Goat Association:,** `https://msfgaregistry.org`
- **Pygora Breeders Association:** `https://pba-pygora.org`

Backpacking with Goats

During the COVID-19 pandemic, people started to change the way they entertained themselves. One thing that changed in the summer of 2020 was outdoor recreation and camping, as people sought out recreation opportunities that allowed them to socially distance. Backpacking with a goat is perfect for this!

When you backpack with goats, they're the ones that do the heavy lifting. The goats' food is right there in the mountains or hills where you're hiking, so they only have to carry your food. And you can go places in the wilderness where you'd never be able to drive.

When you choose the goats you want to train to pack, you need to think about their conformation (whether they're healthy and built to withstand a lot of walking), their breed (you probably want a large one), and their sex (you don't want a buck for obvious reasons, does have other jobs to do, and wethers are ideal).

When you get your goats, you need to purchase gear and start training (see Chapter 8). It also helps to find a pack-goat group in your area or online, to learn what tips they have.

After your goat is well trained, you can take him on his first hike. Make sure not too pack too heavily (no more than 20 percent of his body weight), and don't start with an overly ambitious hike.

Offering Buck Service

I discourage people from buying bucks right off the bat, because they have a very strong odor during breeding season, they need to be housed separately from does, they can become aggressive, and they often are an unnecessary expense for anyone who wants to keep a small herd. As a result, buyers return to me asking where they can get their does bred. If I have the right buck for them and their goals, we breed one of my bucks to their doe(s) — for a price. Hiring out a buck for breeding is called *buck service.*

You can provide buck service in a couple of different ways, and I have used all of them at one time or another:

- **Leasing your buck to another breeder:** When you lease your buck to another breeder, the buck goes to live at that person's farm for the purpose of breeding does. Before you lease a buck to another breeder, make sure you know the living conditions and the health status of her goats, and make sure you have a contract. Contract terms need to cover at least the following points:
  - Period of time the buck will stay
  - Charge for breeding each doe and when that fee will be paid
  - Agreement that the buck will be adequately cared for
  - Requirement that the person leasing will communicate with the buck owner if the buck gets sick or injured
  - Who is responsible if the buck gets sick or dies, including timeliness in contacting a veterinarian, who will pay for veterinary care, and whether additional money is due if he dies
  - What the buck will be fed and who will pay for it
  - How many does the lessee plans to have bred and whether any does from another farm may be bred to the buck
  - Acknowledgment that the buck and does have no known diseases
  - What happens if the does don't *settle* (get pregnant) — most breeders provide a second breeding at no additional cost
  - Agreement that buck owner will sign a *service memo* (a document required by registries to prove parentage)
- **Driveway breeding:** When you breed a goat by bringing the buck or the doe to another farm for only that purpose, it's called a *driveway breeding.* You can do a driveway breeding at your farm or another farm. Usually you bring the

doe to the buck, because he is providing the service. A driveway breeding has the advantages of exposing your buck to much less health risk and letting you control the breeding because the goats spend very little time together.

- **Boarding doe(s) with your buck:** If you agree to board does for breeding with your buck, you need to have a stall where they can be kept separate from the rest of your goats. You can let the buck live with them, or just bring him in when they are in heat. This helps to ensure the safety and health of the doe and the rest of your goats.

  I usually suggest that the breeder bring the doe when he knows she is in heat or a few days before she is due to be in heat (see Chapter 12). If we're lucky, what started out as a boarding becomes a driveway breeding. And if not, I usually keep the doe for a month to ensure that she is with the buck 21 days or so later when she goes into heat again.

WARNING

Make sure that you protect yourself and your goats with a contract. Most of the same terms apply for boarding a doe and leasing a buck. You especially want to make sure that the doe coming to your property is healthy.

Boarding Other People's Goats

Boarding other goats is a good way to make a little extra money and help subsidize your farm if you're set up for it. After I had downsized my herd, I was able to board goats for friends who had purchased new property and had to build a barn before they could move their goats there. Even though their goats were all related to mine (they never kept a buck), we kept them separate to avoid a lot of chaos.

You'll need a separate, secure, fenced area that is not accessible to your herd, along with a safe shelter for them to rest and sleep in. Don't knowingly take in sick goats, unless your intent is to nurse them.

TIP

Make sure both parties agree, at a minimum, on what your responsibilities are, how long you'll board the goats, who will pay for feed, and what to do in case of any health issues. Having it in writing protects you both.

Selling Compost

I think every goat owner can agree: Goats produce a lot of poop. What do you do with all of it? As it turns out, compost made of goat manure and straw (goats' bedding) is great for the garden.

Whether you want to go so far as to develop a system that speeds along the process or just pile up the muck and let nature work its magic, the end result will be the same: rich, black dirt also known as *compost.* Compost is good for the soil because it helps balance pH, retain moisture, prevent compaction, add body, and retain nutrients. Check out *Organic Gardening For Dummies* to get the complete lowdown on compost.

WARNING

You can just put goat manure right on the garden because, unlike chicken and sheep manure, goat manure isn't "hot" — that is, it has a lower concentration of nitrogen, phosphorus, and potassium. However, experts recommend aging manure for at least six months or composting it to 150° Fahrenheit to prevent the spread of diseases such as listeriosis to humans when they eat garden produce that is contaminated.

I trade compost that accumulates in a pile on my property for meat and vegetables from neighboring farmers, because we tend to deal more in barter than money in my neck of the woods. If you don't have a pasture or garden that will benefit from the addition of your excess goat compost, you can probably sell it through word of mouth, by advertising on Craigslist, or by posting signs in local stores and watering holes.

Hiring Out for Weed Control

More and more cities and other municipalities are using goats and sheep to control noxious weeds. This practice has at least two benefits: It provides an alternative to using herbicides, and it creates *firebreaks* (areas cleared of combustible material) that help control wildfires.

You can use your own goats to eat noxious weeds on your property, but if you want to make money and provide your goats with free browse, consider hiring them out. You may find more opportunities to do this as laws restricting the use of pesticides and herbicides get stricter.

To hire out your goats you need:

- Good transportation for getting goats to and from the site
- Portable fencing, so that you keep the goats safe and focused on the project for which they've been hired
- A guardian dog or other protection in areas with predators
- Insurance to cover any mishaps
- A written contract outlining both parties' responsibilities

Talk to other goat owners who have rented out their goats for vegetation control to learn more about other requirements by organizations that hire goat herds, how much you should expect to charge, and contract provisions you need to include before you offer the services of your goats.

Providing "Goat Therapy"

Anyone who has had goats and really gotten to know them will tell you how relaxing they are to be around. Maybe they hypnotize us with the constant, repetitive ruminating. (Not to mention that there's nothing like the hard labor of mucking a barn to leave you feeling relaxed.) I call this calming effect *goat therapy*.

I've had the good fortune to spend a large part of my career working from home, so instead of (or along with) a coffee break I can take a "goat break." After more than 20 years with goats I still enjoy walking to the barn to hang out with these amazing animals.

They're usually calm (when not scuffling for food or in heat or rut) and enjoy being petted. Although I'm not aware of any studies that have been conducted with goats specifically, many studies have shown that having a pet can lower your heart rate and prevent depression. And if you need some exercise, you can take your goat on a walk.

Because of goats' gentle nature, a few people have begun to use them as service animals for children with autism or to visit elderly people who live in nursing homes (see Chapter 8).

When everything else seems to be going wrong, there's nothing like a little goat therapy.

5 The Part of Tens

IN THIS PART . . .

Avoid ten common mistakes new goat owners make.

Dispel ten misconceptions about goats.

Find out how to make delicious goat-milk recipes.

IN THIS CHAPTER

» Getting the right goats

» Making sure you're ready for goats

Chapter 19
Ten Common Mistakes First-Time Goat Owners Make

After more than two decades of raising goats, I have seen people make a lot of mistakes, and I have made some myself. The average length of time that a person owns goats is only around five years. One reason is that they have unrealistic expectations of what owning goats requires.

This chapter covers some of the most common mistakes that people make when they start on their goat adventure. I hope it saves you some time and energy and helps keep you from making mistakes.

Getting Too Many Goats Too Fast

Time and again I have seen people decide to get into goats and then go out and start buying goats from various farms that are reputed for their quality. What they fail to consider is that the goats may not just double in number during the first kidding season — they may triple or quadruple if the goats are fertile! And these new little goats will grow — leading to cramped quarters, more likelihood of disease due to crowding, and more money coming out of your pockets for food and equipment.

I started reasonably with two doe kids, planning to breed them that fall and milk them in the spring when they kidded. I didn't realize how cute their kids would be and what a hard time I would have giving them up. Each doe had a buckling and a doeling that next spring and I just *had* to keep one of each. Within three months I needed a separate space for my buck, which meant getting another buck or a wether. This continued for a number of years (think *rabbits*) until I finally came to my senses and reversed the trend.

So I am speaking from experience when I say: Start slow and don't get the maximum number of goats that your farm can handle right away. And then think about what you are doing when you start breeding your goats. They grow exponentially, and all of the kids are way too cute!

Failing to Educate Yourself before Getting Goats

I don't know how many times I've heard, "I want to get a couple of goats to control the blackberries on my property." When I ask about fencing, where the goats will sleep, or how the potential owners will protect their goats from the things they *don't* want the goats to eat, they seem puzzled. They hadn't thought about that; in fact, they didn't even know that these could be problems.

REMEMBER

Raising goats is a big responsibility. You have these animals' lives in your hands and need to take that seriously. You can't just put them out in a field and expect them to do well. They need to have their hooves trimmed; to be monitored for parasites; to be given a safe shelter, clean water, and appropriate food; and so on. They require time, money, care, and knowledge to thrive and survive.

Not educating yourself about goats and goat ownership is a guarantee for a short-lived goat experience, a lot of grief, or a much more expensive undertaking than you had planned.

Underestimating the Costs

Whether you just want to raise a couple of goats for milk or pets, or a herd of meat goats, don't make the mistake of underestimating the setup and ongoing costs. I've had to add new fencing to my property and replace fencing that had worn out a lot more often than I had anticipated. And who realized that those plastic water buckets actually break after a few years? Or that rats can chew through plastic garbage cans?

One stanchion might work for the two dairy goats you have in the beginning, but down the road you find that you need one or two more. And routine medications actually expire after a year or two and have to be replaced even though you haven't used them.

Raising goats reminds me of what my grandpa used to say: "If I had a million dollars I would go back into farming until the money ran out."

WARNING

Failing to provide for the goats, in the beginning or along the way, will cost more in the long run — whether it be vet bills, losses to predators, the annoyance of goat escapes, or even your back when you have difficulty controlling the goats to do routine care.

Paying Too Much or Too Little for Your Goats

Not all offspring from prize-winning, high-milking, well-muscled, or perfectly conformed goats will be champions, good producers, or even acceptable for anything other than pets.

Buying goats from a breeder who has a good website and sells a lot of goats or one whose goat won a junior championship doesn't guarantee that what they sell you will be the best. Find out the going rate for the type of goat you want, comparison shop, and if you can, ask other buyers about their experiences buying from the seller.

Besides figuring out what the goats are worth to you, consider whether the high-priced goats will better meet your needs or pay for themselves in the long run. On the other hand, beware of bargain goats. There's no such thing as a free goat! If someone offers goats for next to nothing, ask why. You want to get a suitable goat the first time around rather than going through a bunch of cheap goats that don't fit your needs or keeping unsuitable goats because you've gotten attached to them.

Getting Only One Goat

In almost all cases, getting only one goat is a recipe for trouble. Goats are not dogs and do not thrive on human companionship alone.

REMEMBER

Goats are herd animals and need other goats to keep them healthy and happy. A goat without a friend will cry and can even become depressed. Never get just one goat; always buy at least two goats so they can keep each other company.

Dogs, pigs, and cows are not the best friends for a goat. Sheep, rabbits, and horses can be friends with goats. Goats have traditionally been kept with racehorses and are known to have a calming effect on them. Remember that if you do decide to bond your goat with an animal other than another goat, the animal needs to stay with the goat.

People with small farms who have only a few animals sometimes keep a goat and a sheep together. This situation has certain risks:

- The copper needs of sheep and goats are substantially different, and the minerals needed by goats contain a level of copper that is toxic to sheep.
- The only known cases of *scrapie* (a disease like "mad cow disease") in goats were in goats that were housed with sheep. Sheep can also give goats OPPV (the sheep version of CAEV).
- Male sheep (rams) and female goats can and will interbreed, but the fetus normally dies before birth.
- Because goats rear up to fight and sheep charge, goats are more likely to get injured, especially if a buck and ram are intermixed.

Buying Unhealthy Goats

A lot of people are tempted to buy goats at an auction where they can get a "good deal." Although some of these animals may be a good deal, you can't tell by looking at them whether they have an underlying health problem. Many owners of large herds take their goats to auction to get them out of the herd because they have an illness or don't meet their herd requirements, so either don't take that chance or be prepared for a negative outcome.

I am not discounting those among us who rescue injured or sick animals. I am just warning against inadvertently buying a goat that may be sick. If you have other goats, you can bring diseases from the sale barn that affect your herd.

When you buy from a breeder, look closely at the goats and make sure to ask about the goats' health, the herd health, and whether they have been tested for certain diseases. Get a contract so that you have legal recourse if you later find that the goat had a pre-existing health problem.

Neglecting Routine Management and Care

Properly caring for goats is a daily job. In most cases, goats need to be fed and given clean water twice a day. In addition, they require periodic hoof trimming, routine medical care, handling, and grooming. If you let these responsibilities slide, the goat is defenseless, much like a child who relies on you.

Neglecting routine care adds to your workload because you have to work twice as hard to catch up. Worse, you put your goats at risk of poor health when they have to drink from dirty buckets, eat hay off the ground, and walk on overgrown hooves. Don't make this mistake; set a routine that you and your goats can rely on.

Overlooking Your Goats' Dietary Needs

Unless you have a huge acreage that is covered with plentiful weeds and other plants that goats like to eat, you need to buy them hay. Don't expect your goats to spend their time eating grass on a lawn; they like a variety of plants, which is why they're considered browsers rather than grazers.

Goats also need mineral supplementation, which varies according to the part of the country you live in. Different areas are deficient in different essential minerals and you need to provide those in the proper amounts. Dairy goats that you are milking also need grain, alfalfa, and sunflower seeds to keep their production high and their milk rich in butterfat.

You get back what you put into a goat, and even pets or goats whose job is just to clear the property will be healthier if they are fed right. Chapter 6 tells you more about feeding goats.

Giving the Goats Too Little Attention

Although they have been domesticated for over 10,000 years, goats are quick to return to a feral state. This means that they need attention and handling if they're to remain friendly.

Often, goat owners avoid only certain goats — maybe the smelly bucks or the skittish members of the herd that require effort to catch. These goats then get harder and harder to catch for routine care.

You can avoid this mistake and make goat ownership easier on yourself by taking time to interact with your goats several times a day, when you change their water or feed them. At the same time, you can do a daily visual herd check to make sure that they aren't showing any signs of illness.

REMEMBER

If you decide to breed your goats, make sure to handle the kids regularly while they're still small. You can't just let them run with the adults and have no human contact and then expect that they'll easily warm up to you.

Getting a Buck before You're Ready

Bucks are not for everyone. Raising a buck is a big responsibility and an aromatic proposition. Bucks don't normally make good pets and you shouldn't get one with that idea in mind. If you want pets, start with wethers or, if you must, does. They don't stink and are not as strong and aggressive.

TIP

If you're raising goats for milk or meat, ask about leasing a buck from the breeder where you bought your does, or doing a driveway breeding instead of purchasing your own buck. (Chapter 12 tells you more about breeding.) You'll save money in feed and housing, and your property will smell better. You'll also be spared the expense of owning at least one other goat in addition to the buck — a wether — so the buck has a needed friend.

If you start a herd with two does that you want to breed, I can almost guarantee that you will want to keep one of the bucks that is born the next year. If you have already purchased a buck, now you will have two fighting, reeking animals that need a separate pen.

Because bucks are the basis of your breeding program, think it through before you buy your own buck. You won't regret it.

» Knowing the true characteristics of goats

Chapter 20
Ten Misconceptions about Goats

People have a lot of ideas about goats that simply aren't true. As a result, goats have gotten a reputation that they don't deserve. When the people who hold some of these beliefs actually raise goats, the goats can suffer from a lack of proper care. For example, I actually had someone ask me whether they could raise a goat in their house and feed it table scraps — which leads to the first misconception . . . .

Goats Will Eat Anything

No doubt you've seen a cartoon with a goat eating a tin can. That depiction is about as real as a wise-cracking rabbit or a caped crusader. Goats don't eat tin cans. The tin can–eating image of a goat probably came about because a goat was trying to eat the label off of a can. (Yes, they do eat paper, including money.)

Goats are actually quite picky eaters in a lot of ways. They're notorious for wasting hay — after it hits the ground, hay becomes bedding for all but the most starving goat. Goats can be quite difficult about changes in feed until they get used to it or realize that's all they're getting.

Goats do like variety, and they take little nibbles here and there. They also can eat some plants that are considered poisonous to other animals without any problem, but they reject many plants out of hand (or hoof). Goats also are vegetarians, which eliminates a lot of foods from their diet, and they prefer their vegetables raw, thank you very much.

Although goats won't eat just anything, they *will* undoubtedly pull a clean shirt off the clothesline and possibly take a few bites, just like in the song "Bill Grogan's Goat."

Goats Stink

Goats have gotten a reputation for having a strong, disagreeable scent because the bucks do have a powerful odor during breeding season, when they pee on themselves as a kind of cologne for the ladies. They smell pretty mild the rest of the time. The does never have a strong smell, unless they're rubbing up against a buck in love.

Goats create less of a stink in the barn and the barnyard than many other animals — including cattle, horses, and pigs — because of their unique poop. A goat that is healthy and isn't eating too much rich food has nicely pelleted poop with only a mild odor.

Goats Aren't Very Smart

Intelligence in an animal is hard to prove. One of the ways that I gauge the intelligence of an animal is how long it takes to learn its name. My first calf took three months, while most of my goats learn their names in less than a month. I believe that goats' intelligence is on par with that of dogs.

Goats can learn how to open locked gates. They can identify their owners walking two aisles over from their pen at the fair. (I know because my goat Jinx could do just that.) I even had a goat that housebroke herself. Goats can be taught to pack, to pull a cart, and to go through an obstacle course — all indicators that goats are smart.

People who think that goats aren't smart either have them mixed up with sheep or are confusing a goat's refusal to do what they say with stupidity.

Goats Make Good Lawn Mowers

Goats are browsers, not grazers. Goats mow down blackberries and kudzu; sheep mow lawns. Browsers move a lot, preferring leaves, bark, and stems, while grazers stand still to eat vegetation at or near ground level. Browsing gives goats an advantage when snow is covering plants because the parts they prefer are more likely to be exposed. Goats are disadvantaged in terms of nutrition, though, because the parts they like may not have the nutrients that can be found in grass.

Don't expect goats to mow your lawn like sheep would, but if you have pasture they do adapt and learn to use it as a feed source, even though they prefer woody browse.

Goat Milk Tastes Bad

The misconception that goat milk tastes bad is one of the hardest to change, probably because goat milk *can* taste bad. Not following sanitary practices, not straining the hair and hay out of the milk, not cooling it down fast enough, letting it sit too long, and milking near bucks or other odors can cause goat milk to go off flavor.

Goat milk can also develop an "off" flavor if a goat eats a lot of a plant that imparts the odor, such as garlic or onion. In addition, if a doe has *mastitis* (an infection of the udder), it will affect the flavor of her milk. But normally, goat milk is fine.

Fresh goat milk tastes even better than cow milk! The milk from a goat is sweet, and if it has a high butterfat content — such as Nigerian Dwarf milk — it is so rich that it can compete with half and half. I prefer it in my coffee because it adds such a great flavor. If you don't believe me, experiment by getting some fresh goat milk and some cow milk and doing a blind taste test. Even if you can tell the difference, I bet the experiment changes your mind about whether goat milk tastes bad.

Goat Meat Tastes Bad

This misconception goes hand in hand with the belief that goats smell bad. I haven't eaten the meat from a buck, but I have had some of the most tantalizing ribs from a friend's wether. I can assure you that they didn't taste gamey or goaty or anything other than mild and delightful.

Some people are undoubtedly more sensitive to or just dislike goat meat, but I recommend that you try it. (You'll like it!)

Goats Get Most of Their Water from Plants

When I first moved to the country, my neighbor had a buck goat, Odin, that had been given to him as a gift. That neighbor never gave the goat water because he believed that goats get their water from plants and don't need drinking water. Consequently, my daughter and I spent hot summer afternoons taking bowls of water to that poor goat as he stood tethered in the orchard. He rushed to the bowl and greedily drank it.

Goats need a lot of fresh, clean water and can drink more than a gallon a day. It would be hard to get this much water from a plant diet, particularly in a dry area. Even in lush areas a goat can get dehydrated from lack of water. And in males, especially wethers, lack of water can lead to the formation of urinary stones that can get stuck and block the urine, killing the goat.

Don't make the mistake of thinking that you can get by without giving your goats plenty of water (and changing it frequently — not only do goats need water, but they like it clean).

Goats Are Only for People Who Can't Afford Cows

This misconception probably came about because people in developed (or rich) countries are more likely to use cows for milk and meat than those in developing (or poor) countries. But instead of thinking of goats as being for poor people, I think the reality is that goats are for *practical* people.

Goats cost less than cows to feed, need less space, use fewer resources, and are cleaner. They are compact, easy for women and children to handle, and produce just the right amount of milk and meat for a small family. And they are much smarter and more fun than cows!

Only Male Goats Have Beards

Although bucks often have large, impressive beards, a lot of does also grow shorter, more feminine beards. Like horns, goats of both sexes have beards.

TIP

Goats have beards, but sheep don't. Using this knowledge along with the fact that goats hold their tails up and sheeps' tails go down, you can learn to tell the difference between sheep and goats.

People who show their goats normally shave off the female beard and leave on the male beard because they want to accentuate the masculinity or femininity of the animal. This might be one of the reasons that people think only the males have beards — that's what they saw at the fair!

A Dog Makes a Good Friend for a Goat

Goats need other goats. They are herd animals and need the rest of the herd or they will be lonely.

Goats' play is different from dogs' and they don't have the right equipment to defend themselves successfully against a dog's teeth. Horned goats do have some advantage and could injure a dog. Many dogs find sport in chasing a goat — a practice that can cause stress in the goat and even kill it.

My livestock guardian dog, Marley, lived with the bucks and started acting like he was a goat. At one point, the bucks even tried to breed him, so maybe they thought he was a goat, too. But when it came to playing, they weren't evenly matched. They all learned that he could live with them, but he wasn't a friend.

Appendix

Goat-Milk Recipes

RECIPES IN THIS APPENDIX

- Simple Cheese
- Feta Cheese
- Yogurt
- Koumiss
- Goat Milk Chai
- Popovers
- Macaroni and Cheese
- Zucchini Feta Fritters
- Greek Salad
- Lasagne
- Orzo with Spinach and Feta
- Fudge
- Cheesecake
- Pudding

If you have more than a couple of dairy goats and don't have a lot of kids (human or goat), you may find yourself overwhelmed with goat milk. It's great in coffee, on cereal, or just plain, and you can use it to make cheese, yogurt, ice cream, and much more.

Cheeses and Yogurt

Cheeses are satisfying to make and can be simple or complicated. Here are two different cheeses — one that you can make with basic home ingredients and another that requires cheese culture and rennet. Both are easy to make. I also give you a recipe for yogurt, which is another simple way to break into using your goat milk.

TIP

You need the following equipment and supplies to make cheese:

- A 2- or 3-gallon stainless steel pot. (Never use aluminum.)
- Cheese cloth. Buy it at a cheese supply company or as cotton muslin by the yard at a fabric store. Don't use the cheesecloth sold at the grocery store; the weave isn't fine enough.
- A stainless-steel colander.
- A twist tie, rubber band, string, or shoelace to secure the cheese you're hanging to drain.
- A cooking thermometer with a temperature range of 0°F to 212°F.
- A large spoon.
- A long knife for cutting the curds.

Simple Cheese

PREP TIME: ABOUT 15 MIN PLUS ABOUT 3 HR FOR HANGING	COOK TIME: 15 MIN	YIELD: APPROXIMATELY 1 POUND

1 gallon goat milk

¼ cup vinegar or lemon juice

Kosher or any non-iodized salt to taste

1. In a large pot, slowly heat the milk to 195°F.
2. Stir in the vinegar or lemon juice, and keep the milk at 195°F, stirring occasionally, for 10 to 15 minutes. The milk will gradually start to curdle, first with very small curds and then larger curds. When the curds stay separate from the milk when you stir, they're ready.
3. When a soft curd has formed, pour into a cheesecloth-lined colander.
4. Sprinkle the salt over the curd, and then remove the cheese in the cheesecloth and hang it to drain for several hours.
5. When the cheese reaches the desired consistency, add whatever ingredients you like to flavor it (such as salt, herbs, or even fruit and powdered sugar) and refrigerate. The cheese will keep for about 1 week.

PER SERVING: *Calories 271 (From Fat 146); Fat 16g (Saturated 10g); Cholesterol 43mg; Sodium 195mg; Carbohydrate 18g (Dietary Fiber 0g); Protein 14g.*

NOTE: This straightforward recipe is a good place to start your goat-milk recipe exploration.

TIP: Never use ultrapastuerized milk from the store — it won't turn into cheese.

Feta Cheese

PREP TIME: ABOUT 10 MIN PLUS ABOUT 32 HR FOR HANGING	COOK TIME: 75 MIN	YIELD: 2 POUNDS

2 gallons goat milk

¼ teaspoon mesophilic culture (see Tip)

¼ teaspoon liquid vegetable rennet or ½ rennet tablet

¼ cup cool water

Kosher or any non-iodized salt to taste

1. In a large pot, warm the milk to 86°F and stir in the mesophilic culture. Let the mixture ripen for 1 hour.
2. In a measuring cup, mix the rennet into the cool water and stir the mixture into the milk. Cover and allow the mixture to set for another hour to coagulate.
3. Using a knife that reaches the bottom of the pot, slice across the cheese in lines from left to right, spaced ½ inch apart. Turn the pot 90 degrees and cut across the previous cuts, in lines spaced ½ inch apart. Then slice at an angle through the cheese at ½-inch intervals until the cheese is cut into cubes. Allow to rest for 5 minutes.
4. Stir gently for 15 minutes, keeping the curds at 86°F. Cut any pieces that were missed so that the pieces are uniform in size.
5. Pour the curds into a cheesecloth-lined colander. You can save the whey for ricotta cheese or give to chickens, pigs, or dogs.
6. Tie the bag of curds and hang it to drain for 4 to 6 hours or overnight.
7. Remove the cheese from the cheesecloth, slice it in half, and lay the cheese in a bowl.
8. Salt all the surfaces, cover with a plate or plastic wrap, and allow it to set at room temperature for 24 hours. During the 24 hours, intermittently drain the whey that is expelled from the cheese and resalt the cheese.

9 Put the cheese in a covered container and refrigerate. The cheese will keep for 5 to 7 days.

PER SERVING: *Calories 270 (From Fat 146); Fat 16g (Saturated 10g); Cholesterol 43mg; Sodium 958mg; Carbohydrate 17g (Dietary Fiber 0g); Protein 14g.*

TIP: Mesophilic culture is a culture that is destroyed at higher temperatures. You can get it at a cheese supply store. You can get rennet at a cheese supply store, too.

NOTE: The beauty of this cheese is that it's very forgiving — if you add the rennet late, cut the curds late, or let it hang or expel whey for a longer time than the recipe calls for, it will still turn out fine. You also can't oversalt it because it's a salty cheese and the salt helps remove the whey.

Yogurt

PREP TIME: 15 MIN PLUS 8 HR FOR SETTING | YIELD: 2 QUARTS

½ gallon goat milk

1 to 2 tablespoons plain yogurt

1. In a large pot, heat the milk to 110°F.
2. Stir in 1 or 2 tablespoons of plain yogurt with a wire whisk.
3. Pour the mixture into sterile quart or pint jars and place in a cooler with a jar of very hot water.
4. Cover everything with a towel, close the cooler, and wait 8 hours. When you open the cooler the next morning — *voilà* — yogurt! Use the last bit of each batch to start the next one.

PER SERVING: *Calories 85 (From Fat 46); Fat 5g (Saturated 3g); Cholesterol 14mg; Sodium 61mg; Carbohydrate 6g (Dietary Fiber 0g); Protein 4g.*

NOTE: Yogurt is a fermented milk product that contains friendly bacteria to aid the digestive system. It's good for goat owners *and* their goats. If you have a goat on antibiotics for any reason and you don't have a probiotic on hand, give the goat a couple tablespoons of yogurt to replace the good bacteria that the antibiotic kills.

TIP: Homemade goat milk yogurt doesn't turn out as thick as the grocery-store yogurt you're used to, so don't be put off thinking you made a mistake. It makes great smoothies!

TIP: Follow these steps in the evening for yogurt the next morning.

Drinks

I'm a big fan of drinking goat milk straight, but you also can turn it into tasty beverages like koumiss, a fermented, carbonated dairy drink that was traditionally made from mares' milk in Central Asia.

Koumiss

PREP TIME: 30 MIN PLUS 24 HR SITTING TIME AND 24 HR STORING TIME BEFORE REFRIGERATION	COOK TIME: 30 MIN	YIELD: ABOUT 1 GALLON

1 gallon goat milk

2 tablespoons honey

¼ cup water

⅛ teaspoon champagne yeast

1. In a stainless-steel pot, heat the milk to 180°F; remove any film that forms.
2. Add the honey and allow the mixture to cool to 70°F.
3. Heat the water to 115°F, dissolve the champagne yeast in the water, and let the mixture stand for 10 minutes.
4. Stir the yeast mixture into the milk. Cover with a clean cloth and allow to stand at room temperature until it foams, about 24 hours.
5. Pour the koumiss into sanitized beer bottles that can withstand carbonation pressure. Fill only to 1 inch below the bottom of the neck of the bottle.
6. Store at room temperature for 24 hours, and then refrigerate. Shake the bottles *gently* every few days but not just before opening.

PER SERVING: *Calories 177 (From Fat 91); Fat 10g (Saturated 7g); Cholesterol 27mg; Sodium 122mg; Carbohydrate 13g (Dietary Fiber 0g); Protein 9g.*

TIP: You can buy champagne yeast at a beer makers' shop.

TIP: Koumiss will keep for 6 to 8 weeks but then becomes increasingly acidic. You can add honey to hide the acidic taste.

TIP: Before you start, sanitize everything that will come into contact with the drink mixture, using 1 tablespoon of bleach to 1 gallon of water. Soak all bottles, containers, and utensils for 15 minutes in the bleach mixture; then rinse in hot water, making sure to remove all traces of bleach.

Goat Milk Chai

PREP TIME: 15 MIN | COOK TIME: 15 MIN | YIELD: 4 SERVINGS

4 cups water

¼ cup fresh grated ginger root

One 1-inch piece of cinnamon stick

7 black peppercorns

7 white peppercorns

7 red peppercorns

7 allspice berries

7 whole cloves

15 cardamom pods

3 cups goat milk

½ cup maple syrup

3 black tea bags (or more, if you like stronger tea)

1. In a 2-quart pot, place the water and ginger and bring to a boil.
2. While waiting for the water to boil, grind the cinnamon, peppercorns, allspice, cloves, and cardamom with a mortar and pestle or electric grinder. Add to the water.
3. Decrease the heat and simmer for 15 minutes.
4. Add the goat milk, maple syrup, and tea. Bring to a boil and immediately remove from the heat.
5. Strain and serve.

PER SERVING: *Calories 135 (From Fat 40); Fat 4g (Saturated 3g); Cholesterol 12mg; Sodium 55mg; Carbohydrate 21g (Dietary Fiber 0g); Protein 4g.*

NOTE: *Chai* (pronounce to rhyme with *tie*) simply means *tea,* but it usually refers to a spiced version with milk along the lines of the one in this recipe.

Entrees and Side Dishes

After you've experimented with goat-milk cheese, you're probably ready to dive into further dishes that put it to use. Here are some of the recipes my family loves.

Popovers

PREP TIME: 10 MIN	COOK TIME: 30 MIN	YIELD: 1 DOZEN

4 large eggs

2 cups goat milk

2 cups flour

1 teaspoon salt

1. Preheat the oven to 450°F.
2. Grease a popover pan or muffin tin and dust it with flour so the batter will have something to climb as it rises.
3. In a large bowl, beat the eggs slightly with a wire whisk.
4. Add the milk, flour, and salt. Beat until smooth. Do not overbeat.
5. Fill the muffin cups two-thirds full.
6. Bake for 20 minutes and then lower the oven temperature to 350°F.
7. Continue baking until the popovers are well browned and crusty, about 10 minutes more. (If the popovers aren't brown enough, they'll collapse when they cool.)
8. Remove from the oven and serve immediately.

PER SERVING: *Calories 172 (From Fat 43); Fat 5g (Saturated 2g); Cholesterol 100mg; Sodium 314mg; Carbohydrate 24g (Dietary Fiber 1g); Protein 8g.*

TIP: Avoid opening the oven until the last 5 minutes to prevent the popovers from deflating.

NOTE: Savory popovers are great for breakfast or alongside dinner.

Macaroni and Cheese

PREP TIME: 10 MIN	COOK TIME: 35–45 MIN	YIELD: 4 SERVINGS

8 ounces macaroni

1 egg, beaten slightly

2 cups goat cottage cheese

1 cup sour cream

8 ounces goat cheddar cheese

Salt to taste

1 teaspoon paprika

1. Preheat the oven to 350°F.
2. In a pot filled with water, cook the macaroni for about 5 minutes.
3. While the macaroni is cooking, combine the egg, cottage cheese, sour cream, cheddar cheese, and salt in a large bowl.
4. Drain the macaroni.
5. Add the drained macaroni to the bowl with the cheeses, and pour the mixture into a buttered 9-x-13-inch baking dish.
6. Sprinkle the top with paprika and bake for 30 to 45 minutes, or until lightly browned.

PER SERVING: *Calories 355 (From Fat 201); Fat 0g (Saturated 12g); Cholesterol 92mg; Sodium 904mg; Carbohydrate 16g (Dietary Fiber 1g); Protein 24g.*

Zucchini Feta Fritters

PREP TIME: 40 MIN | COOK TIME: 30 MIN | YIELD: 4 SERVINGS

1 pound coarsely grated zucchini

Salt to taste

½ pound goat feta cheese, crumbled

1 cup minced herbs (parsley, mint, dill, basil, or your choice of any combination)

3 slightly beaten eggs

Pepper to taste

6 chopped green onions or ½ onion

1 cup flour

Vegetable oil for frying

1. Grate the zucchini into a colander. Salt it and let sit for about 30 minutes.
2. Squeeze out the excess moisture.
3. Transfer the zucchini to a bowl, and mix in the cheese, herbs, eggs, salt, pepper, and onions, mixing well.
4. Add the flour and mix well again.
5. In a large frying pan, heat the oil. Drop the batter into the pan about ¼ cup at a time.
6. Fry both sides, about 3 minutes per side.
7. Remove and drain on paper towels.

PER SERVING: *Calories 483 (From Fat 312); Fat 35g (Saturated 5g); Cholesterol 166mg; Sodium 201mg; Carbohydrate 32g (Dietary Fiber 3g); Protein 12g.*

VARY IT! These zingy fritters can be seasoned however you like. Add a pinch of crushed red pepper to up the spice ante.

Greek Salad

PREP TIME: 5 MIN | YIELD: 4 SERVINGS

2 red bell peppers, chopped

1 pound goat feta cheese, cut into cubes

1 cup Greek olives, pitted

½ cup olive oil

1 small red onion, diced

2 teaspoons minced garlic

1 tablespoon balsamic vinegar

2 teaspoons fresh thyme or 1 teaspoon dried thyme

Salt to taste

Pepper to taste

Lemon juice to taste

1 In a large bowl, combine the bell pepper, cheese, and olives.

2 Add the olive oil and toss gently.

3 Add the remaining ingredients and toss gently again.

4 Adjust the seasonings, cover, and chill at least 5 hours before serving.

PER SERVING: *Calories 437 (From Fat 358); Fat 40g (Saturated 8g); Cholesterol 12mg; Sodium 914mg; Carbohydrate 17g (Dietary Fiber 2g); Protein 5g.*

TIP: Toss cucumbers and tomatoes into this salad if you have them on hand.

NOTE: The recipe makes plenty of dressing, so you may have some left over for another meal.

Lasagne

PREP TIME: 15 MIN | COOK TIME: 1 HR 10 MIN | YIELD: 8–10 SERVINGS

8 ounces lasagna noodles

2 medium onions, chopped

4 cloves garlic, minced

3 tablespoons olive oil

2 cups canned tomatoes

2 teaspoons dried oregano

1 teaspoon dried basil

¼ cup fresh chopped parsley

2 teaspoons salt

¾ pound sliced mushrooms

One 15.5-ounce can pinto beans (or ¾ cup dried, cooked tender), drained and rinsed

¾ pound goat mozzarella cheese, divided

2 cups goat cottage cheese, divided

½ cup fresh grated Parmesan cheese, divided

1. Preheat the oven to 375°F.
2. In a pot filled with water, cook the lasagna noodles for 10 minutes. Rinse in cold water to prevent sticking; set aside.
3. In a frying pan, sauté the onions and garlic in the olive oil until soft and transparent but not browned.
4. Stir in the tomatoes, oregano, basil, parsley, and salt. Cook for 30 minutes, simmering and stirring often until it thickens.
5. Stir in the mushrooms and beans.
6. In the bottom of a 9-x-13-inch baking dish, place a layer of one-third of the noodles.
7. Spoon one-third of the sauce over the noodles.
8. Spread 1/4 cup of the mozzarella cheese, 2/3 cup of the cottage cheese, and 8 teaspoons of Parmesan cheese over the noodles.
9. Place another layer of one-third of the noodles in the dish.
10. Repeat Steps 6 through 8.
11. Repeat Steps 6 and 7.
12. Bake for 35 to 40 minutes or until lightly browned.
13. Cool for 5 to 10 minutes before cutting and serving.

PER SERVING: *Calories 429 (From Fat 166); Fat 19g (Saturated 12g); Cholesterol 51mg; Sodium 1249mg; Carbohydrate 37g (Dietary Fiber 5g); Protein 30g.*

NOTE: So easy to prepare, this lasagne adds beans for a vegetarian twist and keeps well in the fridge or freezer.

Orzo with Spinach and Feta

PREP TIME: 15 MIN	COOK TIME: 20 MIN	YIELD: 6 SERVINGS

½ pound (1¼ cups) orzo pasta

2 tablespoons olive oil

2 tablespoons butter

3 cloves garlic, minced (about 1 tablespoon), or to taste

¼ teaspoon dried hot red pepper flakes

3 green onions, chopped

¾ pound fresh spinach, chopped

½ cup crumbled goat feta cheese

Salt to taste

Pepper to taste

1. In an uncovered 4- to 6-quart pot of boiling salted water, cook the pasta, stirring occasionally, until al dente, about 5 minutes.
2. While the pasta is cooking, heat the oil and butter in a heavy skillet over medium-high heat until hot but not smoking.
3. Sauté the garlic, red pepper flakes, and onions, stirring occasionally, until the garlic is golden, about 2 minutes.
4. Add the spinach and cook, stirring for about 2 minutes or until completely wilted.
5. Drain the pasta, transfer it to a bowl, and cover to keep warm.
6. Toss the pasta with the spinach mixture and the cheese until combined.
7. Season with salt and pepper.

PER SERVING: *Calories 245 (From Fat 83); Fat 9g (Saturated 3g); Cholesterol 11mg; Sodium 117mg; Carbohydrate 36g (Dietary Fiber 4g); Protein 7g.*

NOTE: Orzo is quick-cooking, rice-shaped pasta.

Desserts

Add just a few milking goats to your herd, and you may feel you have milk coming out your ears. Fortunately, goat milk works as well for sweet treats as it does for savory dishes.

Fudge

PREP TIME: 15 MIN PLUS 1 HR FOR CHILLING | YIELD: 24 SERVINGS

3 cups sugar

⅔ cup goat milk

¾ cup margarine

6 ounces chocolate chips

7 ounces marshmallow creme

1 teaspoon vanilla flavoring

1. In a heavy 1-quart saucepan, combine the sugar, milk, and margarine. Slowly bring to a full rolling boil, stirring constantly.
2. Reduce the heat to medium and continue boiling for 5 minutes.
3. Remove from the heat and stir in the chocolate chips until melted.
4. Add the marshmallow creme and vanilla, beating until well blended.
5. Pour the mixture into a greased 9-x-13-inch pan.
6. Cool at room temperature.

PER SERVING: *Calories 213 (From Fat 73); Fat 8g (Saturated 3g); Cholesterol 1mg; Sodium 75mg; Carbohydrate 36g (Dietary Fiber 0g); Protein 1g.*

VARY IT! If you like, add ¾ to 1 cup nuts in Step 4.

NOTE: This very easy fudge recipe gives great results without a lot of work.

Cheesecake

PREP TIME: 20 MIN | COOK TIME: 1 HR | YIELD: 16 SERVINGS

2 cups graham cracker crumbs or chocolate wafer crumbs

1 teaspoon cinnamon or nutmeg (optional)

⅓ cup melted butter

4 eggs

2 cups Simple Cheese (see recipe earlier in this appendix)

1 tablespoon lemon juice

½ teaspoon plus 1 tablespoon vanilla, divided

1⅔ cups sugar, divided

4 tablespoons flour

½ teaspoon salt

2 cups sour cream

1. Preheat the oven to 325°F.
2. In a blender, blend the graham cracker crumbs or chocolate wafer crumbs with the cinnamon or nutmeg and the melted butter.
3. Pour the mixture into a buttered 8- or 9-inch springform pan and press firmly onto the bottom and up the sides approximately 1 to 2 inches.
4. In a blender, blend the eggs, cheese, lemon juice, and 1/2 teaspoon of the vanilla until smooth.
5. In a bowl, mix 1 cup of the sugar, the flour, and the salt. Stir this mixture into the egg and cheese mixture.
6. Pour the filling carefully into the crust and bake for 45 minutes to 1 hour, or until lightly browned. The filling should be somewhat set.
7. In a bowl, mix the sour cream, the remaining 2/3 cup of sugar, and the remaining 1 tablespoon of vanilla to make the topping.
8. Pour the topping over the hot cheesecake and bake an additional 10 minutes.
9. Cool at room temperature for several hours and then in the refrigerator for 4 hours or overnight.
10. Remove the sides of the pan and serve.

PER SERVING: *Calories 515 (From Fat 228); Fat 25g (Saturated 14g); Cholesterol 155mg; Sodium 350mg; Carbohydrate 65g (Dietary Fiber 1g); Protein 8g.*

Pudding

PREP TIME: 5 MIN	COOK TIME: 5 MIN	YIELD: 4 SERVINGS

¾ cup sugar

4 egg yolks

2 tablespoons flour

2 cups goat milk

1 teaspoon vanilla

1. In a pot over medium heat, stir the sugar, egg yolks, flour, and milk until just boiling.
2. Reduce the heat and simmer for 5 minutes. (For a thicker pudding, continue to simmer while stirring for another few minutes.)
3. Remove from the heat, stir in the vanilla, and let the mixture sit for a few minutes while it thickens, or pour into smaller cups and chill for at least 5 hours.

PER SERVING: *Calories 305 (From Fat 92); Fat 10g (Saturated 5g); Cholesterol 226mg; Sodium 69mg; Carbohydrate 46g (Dietary Fiber 0g); Protein 8g.*

TIP: Eat plain or top with cinnamon, bananas, or fruit.

NOTE: My family likes this one warm, but you can also chill this pudding before eating.

Index

C

D

E

F

G

H

I

J

K

L

M

N

O

P

Q

R

U

V

W

Y

Z

About the Author

Cheryl K. Smith has raised a small herd of Nigerian Dwarf and Oberian dairy goats under the herd name Mystic Acres since 1998. She is the owner of karmadillo Press and is the author of *Goat Health Care*, 2nd Edition; *Goat Midwifery; The Best of Ruminations Goat Milk and Cheese Recipes;* and *Raising Goats: Some Essentials.*

Dedication

This book is dedicated to the memory of Bob Kimball, my farming partner for ten years, who made it possible for me to focus on writing the first edition of this book. He built most of the structures in the book and helped me translate the steps into writing. He also kept the farm going and the goats fed and happy while I gave most of my attention to this project. I couldn't have done it without him.

Author's Acknowledgments

I would never have gotten goats had it not been for Simone Delaty, who showed me her goats and shared her goat cheese with me one evening in 1988 on her farm in Wellman, Iowa. Farming was already in my blood (both of my parents came from Iowa farms), but that event led me to the path I would ultimately take.

I have been fortunate during my years as a goat owner to have a great community of goat owners who share their knowledge and experience with me. My first mentor, Stacy Morris, deserves a special thank-you for sharing so much of her wisdom as I was learning about goats. She also taught me to mentor others, as she so aptly did for me.

I want to give a special thank-you to Sue Weaver, a fellow goat author and wise woman, who served as technical editor and helped make this edition even better than the first, and an essential read for new goat owners. In addition, I want to thank my good friend Hunk-Eye for technical assistance on building a milk stand.

As always, the Wiley staff were wonderful to work with, ensuring that this edition will be another must-read. I especially want to thank Kelsey Baird for coordinating things, Elizabeth Kuball for being a great editor to work with, and Barbara Frake and Kathryn Born for the artwork.

Publisher's Acknowledgments

Acquisitions Editor: Kelsey Baird

Project Editor: Elizabeth Kuball

Copy Editor: Elizabeth Kuball

Technical Editor: Sue Weaver

Proofreader: Debbye Butler

Production Editor: Mohammed Zafar Ali

Illustrator: Kathryn Born, Barbara Frake

Cover Image: © Dudarev Mikhail/Shutterstock